TRACES
OF THE PAST

C H E M I C A L

	Americas	Europe	Africa
B.C. 1,000,000			fire
100,000		cooking, mineral pigments	
35,000			
20,000		baked clay	
12,000	mineral pigments		
9,000			
7,000			
5,000	lime		gold
4,000	worked copper		glaze
3,000			papyrus
2,500	pottery		synthetic pigments
2,000			glass
1,600	gold		enamel
1,200			
1,000	organic dyes, smelting, silver		
800			
600			iron
400			
200			
A.D. 0			mosaic, glass
200	tumbaga, lead		
400			
600			
800			
1,000	bronze, platinum		

T I M E L I N E

West/Central Asia	East Asia
cooking	
adhesives	
stone annealing, baked clay, leather	
lime plaster	
worked copper, pottery	pottery
asphalt, pitch, linen	
smelting, beer, wine, felt, bread, lead, gold	
bronze, glazed quartz, silver	
soap	
iron	silk
glass, lead glaze, mordant dyes	bronze
tin, mercury	
steel, organic dyes	
	lacquer
	lead glass, cast iron
	brass
brass, parchment	
blown glass, cameo glass	zinc
	porcelain, paper, China ink
tin glaze	

	Americas	Europe
B.C. 20,000		
		Gravettian
18,000		Solutrean
15,000		Magdalenian
12,000	Clovis*	
8,000	Archaic*†	
6,000	Middle Preceramic‡	
4,000		Vinca†
3,000	Old Copper Culture* Late Preceramic‡	Troy I*
2,500		Early Helladic/Minoan/Cycladic*
2,000	Preclassic†	
1,800		Middle Helladic/Minoan/Cycladic*
1,600		Mycenean*
1,400	Olmec†	Late Helladic/Minoan*
1,200		
1,000	Chavín‡	Halstatt A,† Geometric*
800		Halstatt B,† Archaic*
600	Early Woodland,* Paracas-Nazca,‡ Zapotec†	Hallstatt C-D,† Classical*
400		La Tène†
200	Middle Woodland,* Mochica,‡ Anasazi*	Hellenistic,* Republican Rome
A.D. 0		Imperial Rome
200	Late Woodland,* Classic,† Maya,† Teotihuácan†	
500		Middle/Dark Ages, Merovingian
800	Toltec†	Holy Roman Empire
1,000	Postclassic,† Chimú,‡ Mississippian*	
1,200	Aztec†	Renaissance
1,400	Inca‡	

* North America	* Mediterranean
† Mesoamerica	† Central
‡ South America	

T I M E L I N E

Africa	West/Central Asia	East Asia
Upper Paleolithic		
Mesolithic		
Neolithic		Jōmon*
Eneo-Chalcolithic		
Bronze Age		
Old Kingdom*	Sumerian*	
	Mohenjo Daro/Harappa†	
Middle Kingdom*	Akkadian,* Old Babylonian*	Bronze Age†
	Old Assyrian*	Shang†
New Kingdom*	Hittite*	
Iron Age		
		Zhou†
	Phrygian,* Neo-Babylonian,* Neo-Assyrian*	
Iron Age†	Scythian†	Iron Age†
Hellenistic*	Hellenistic*	Yayoi,* Quin,† Han†
Imperial Rome*	Imperial Rome*	
	Sasanian*	Warring States,† Yamato/Kofun*
Ummayad Caliphate*	Ummayad Caliphate*	Sui,† Tang,† Nara*
Abbasid Caliphate*	Abbasid Caliphate*	Heian,* Song†
Mamluk Sultanate,* Benin†	Mamluk Sultanate*	Yuan†
Zimbabwe†		Ming†

* North Africa	* Southwest Asia	* Japanese Period
† Sub-Sahara	† Central Asia	† Chinese Dynasty

TRACES
OF THE PAST

•

Unraveling
THE
Secrets
OF
Archaeology
THROUGH
Chemistry

•

JOSEPH B. LAMBERT

Helix Books
Addison-Wesley
Reading, Massachusetts

Many of the designations used by manufacturers and sellers to distinguish their products are claimed as trademarks. Where those designations appear in this book and Addison-Wesley was aware of a trademark claim, the designations have been printed in initial capital letters.

Library of Congress Cataloging-in-Publication Data
Lambert, Joseph B.
    Traces of the past : unraveling the secrets of archaeology through
chemistry / Joseph B. Lambert.
        p.   cm. — (Helix books)
    Includes bibliographical references and index.
    ISBN 0-201-40928-3
    1. Archaeology—Methodology.   2. Archaeological chemistry.
I. Title.
CC75.L297   1997   97–11454
930.1'028—dc21   CIP

Addison-Wesley is an imprint of Addison Wesley Longman, Inc.

Text design by Diane Levy
Set in 11-point Garamond by Argosy

123456789—MA—0100999897
First printing, August 1997

Find us on the World Wide Web at
http://www.aw.com/gb/

C o n t e n t s

Preface

Over the last few decades a hitherto unappreciated common ground between chemistry and archaeology has begun to be explored. Chemistry is the study of matter and its changes; archaeology uncovers and studies the material remains of past societies. The chemist who analyzes these remains has now become partly an archaeologist. The results of such studies provide a better understanding of both fields, not only contributing to our knowledge of past human life but also documenting the early development of chemistry. In this way we can better understand the chemical inventions that have helped move the human race along its cultural evolution. The two themes that are explored in this book are the chemical development of the materials of daily life—from pottery to metals and from beer to paper—and the chemical analysis of these materials as discovered in the archaeological context. These two themes describe chemistry both as creator and as analyst of culture.

Each chapter of this book focuses on a different class of material available to the archaeologist for study. The story begins with stone, the simplest material exploited by humans, and continues with materials that required increasingly complex chemical manipulation to produce: pottery, glaze, glass, and metals from the inorganic world; dyes, food, beverages, clothing, and adhesives from the organic world. From these analyses many questions arise, and some even are answered: Where did the bluestones of Stonehenge come from? How did the ancient Greeks produce both black and red areas on the surface of a ceramic? How does porcelain differ from other ceramics? What was the first synthetic chemical? What were the original contents of an ancient vessel? Did lead poisoning hasten the decline of the Roman Empire? When did humans in North America begin to cultivate and eat maize? When did our modern human species populate the globe? The approaches and answers to these questions illustrate how chemistry has been a constant companion in the development of civilization.

A c k n o w l e d g m e n t s

Each of my students in archaeological chemistry has inspired me in a unique way and then moved on to new chemical horizons. Chuck McLaughlin is a sculptor; Carole Szpunar, a coal and environmental chemist; Sharon Vlasak Simpson, a photographic and imaging chemist; Liang Xue, a nuclear magnetic resonance spectroscopist; Jane Homeyer, the director of a forensic laboratory; and Suzanne Johnson, a food chemist. This manuscript was read in its entirety and improved by Heather Mimnaugh, Mary Lambert, Robert Tykot, Janet Biehl, and Catherine Shawl. Carol Slingo provided expert word processing and many valuable suggestions. Numerous archaeologists and museum curators have assisted me by providing samples for analysis and help on the archaeological side, but none more than Jane Buikstra.

TRACES
OF THE PAST

1

STONE

Elemental Fingerprinting: Quartzite

Imagine a block of stone more than four stories high, almost as deep, and 15 feet wide: a monolith weighing about 720 tons. Thirty-two hundred years ago, during the Eighteenth Egyptian Dynasty, two such blocks were mined in a quarry, taken to Thebes (near present-day Luxor), carved into giant memorials for the pharaoh Amenhotep III, and set up on the west bank of the Nile River (Figure 1.1). As Egyptian fortunes waned, these Colossi of Memnon, as they now are called, were visited by invading armies such as the Assyrians in the seventh century B.C. and the Persians in the sixth and fifth centuries B.C. These armies did their best to damage the Colossi, possibly by building fires around them to crack the stone. The Colossi, however, were made of quartzite, one of the hardest rocks worked in antiquity, so that they survived essentially intact until about 27 B.C., when an earthquake struck. Although the southern Colossus was not harmed, the northern one lost the upper half of his body. According to the Greek historian Strabo, the lower half remained in place. Some two centuries later the Roman emperor Septimius Severus had the damaged Colossus repaired.

The source of these monolithic blocks of stone has been a puzzle ever since antiquity. Quartzite is made of the same substance as quartz sand, silicon dioxide. Mixed with this main constituent are very small amounts of numerous chemical elements other than silicon and oxygen. During the 1970s a group of scientists led by R. F. Heizer found the solution to the puzzle by analyzing patterns in these trace elements. They took samples from several quartzite quarries along the Nile River Valley and compared them with samples from the Colossi. Their objective was to use the relative amounts of these trace elements as a fingerprint to identify the quarry source of the stone. Certain elements are more common in one quarry than in others, and these variations

Figure 1.1. The Colossi of Memnon: a, *the southern Colossus is essentially intact;* b, *the northern Colossus was damaged by an earthquake about 27* B.C., *at least from the waist up, and was later repaired.*

can be used to characterize the source. To the eye, one quartzite may look much like another, but chemical analysis provides distinctions. The elements iron and europium were particularly good at distinguishing the various quarries. Somewhat surprisingly, the giant blocks of quartzite matched the elemental profile of a quarry called Gebel el Ahmar, far downriver near Cairo, rather than anything close to Thebes (Figure 1.2). The stone blocks had been transported upstream.

It is possible that closer quarries were not known during the Eighteenth Dynasty, or maybe quartzite from Gebel el Ahmar was considered to be superior in hardness or appearance. Economic, political, or other considerations also may have contributed to the choice. Whatever the reasons, they had disappeared by the time Septimius Severus came on the scene, as the blocks used by the Romans for the reconstruction contain trace elements that correspond to quarries near Aswan. These quarries are not only closer to Thebes but upstream, so that the blocks presumably could be floated downstream.

A specific choice of raw material had been made by the ancient stonemasons to fulfill a particular function. In this case quartzite was the material of choice because it was available in very large pieces and could be carved as a

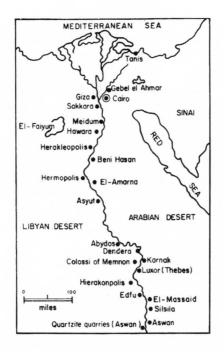

Figure 1.2. The Nile River portion of Egypt. The source of the quartzite used to construct the Colossi, Gebel el Ahmar, is seen just to the north of Cairo. The Aswan quarries, used for the Roman reconstruction, are seen at the southern end of the map.

suitable memorial. For the sake of later generations, it also proved to be extremely hardy. Not only have the Colossi themselves survived to the present day, albeit with some reconstruction, but the hieroglyphics on the original portions are still readable. The ancient engineers clearly made a shrewd choice. They were unaware, however, of those trace elements that would serve as a fingerprint for a later generation to identify the quarry from which the stone had been mined.

Isotopic Fingerprinting: Marble

Chemical analysis can provide information about sources not only of monoliths such as the Colossi, but also of smaller objects such as marble statues. In countless museums throughout the world we can gaze at the countenances of ancient personages, sometimes on a bust, sometimes on a full-standing figure. It is remarkable that such objects have survived the millennia, and in fact much of what we see often is reconstruction. Harvard's Fogg Museum has a very realistic bust said to be that of Antonia Minor (Figure 1.3). Antonia was the daughter of Marc Antony and the mother of Germanicus, who died before he could become emperor, and of Claudius, who indeed became emperor. The bust was part of the collection of Wilton House in England as early as the seventeenth century, but its history before that time is unknown. Visual examination clearly shows that it is composed of five separate parts: the head, the lower part of the hair, and three pieces that comprise the bust (labeled I–V in Figure 1.3). Given the large number of pieces of this marble puzzle, it is reasonable to ask whether they were parts of a single original whole.

Elemental fingerprinting, as was used for the Colossi of Memnon, might be applied to this problem, but Harmon and Valerie Craig developed another chemical approach for identifying sources of classical marble. They examined only two elements, carbon and oxygen, which are major components of the calcium carbonate ($CaCO_3$) that largely makes up marble. What distinguishes one element from all others is the number of protons in its nucleus. The rest of the nucleus is made up of neutrons. For some elements, such as phospho-

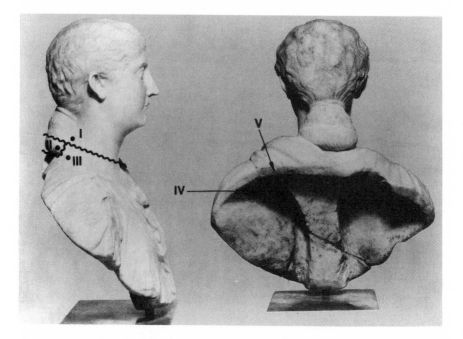

Figure 1.3. A Roman bust of Antonia Minor. The five pieces are labeled I–V. Isotopic analysis indicates that the head is of Parian marble, while the hair bun is of Italian Carrara marble. The three portions of the bust are from Carrara (one piece) and Paros (two pieces).

rus, only one stable form occurs in nature, but for others, such as carbon and oxygen, variations in the number of neutrons produce different elemental forms, or *isotopes*. As long as there are exactly six protons, the element is carbon. The most common isotope of carbon is carbon-12, but in addition there are carbon-13 and carbon-14 (the latter is unstable and hence radioactive). The numbers 12, 13, and 14 provide a count of the total number of heavy particles inside the carbon nucleus. As each has six protons, they respectively have six, seven, and eight neutrons. (Carbon isotopes reappear at the end of this chapter, in a discussion of dating methods.) Oxygen occurs primarily as oxygen-16, but also as oxygen-17 and oxygen-18. All the oxygen isotopes are indefinitely stable, that is, they are not radioactive.

The exact proportions of the various isotopes can vary from place to place, because their slight differences in weight result in biological and geological fractionation. Imagine a stream running over rocks, pebbles, and sand. The smaller particles move downstream more rapidly. Similarly, atmospheric, geological, or biological processes (wind, water moving through soil, metabolism) can transport substances containing light isotopes faster than those with heavy isotopes. Thus the raw material from which rocks and stones are made (water provides oxygen, carbon dioxide provides carbon) can have different proportions of the carbon and oxygen isotopes, depending on local conditions. At the time a stone is formed geologically, the isotopic proportions are sealed in. Isotopic variations from place to place thus can provide another fingerprint of the source, or *provenance,* of the stone.

The Craigs found that the most famous classical sources of marble have distinct oxygen and carbon isotopic proportions. On this basis, actual works of art can be attributed to specific quarry sources. The earliest Greek statuary seems to have been made of marble from the Cycladic Islands, which include Naxos and Paros (Figure 1.4). By the sixth century B.C., Athenians had discovered marble in nearby Mount Pentelikon, from which the Parthenon was largely constructed. From the fifth century B.C. into Roman times, marble from Mount Hymettos was widely used, but sources in Italy and Turkey also increased in popularity. To put Antonia back together, it was necessary to examine the isotopic signatures of all feasible sources right up to the Renaissance. Norman Herz carried out such a study and concluded that probably only the head, which proved to be of Parian marble, was authentic. The carefully matched lower part of the hair, however, was of Italian Carrara marble, as was one of the pieces of the bust. The other two bust fragments also were Parian, but had isotopic signatures clearly different from the head. Thus the Wilton/Fogg Antonia had been reconstructed at least once and possibly several times, with the addition of pieces of a classical bust and of Italian marble.

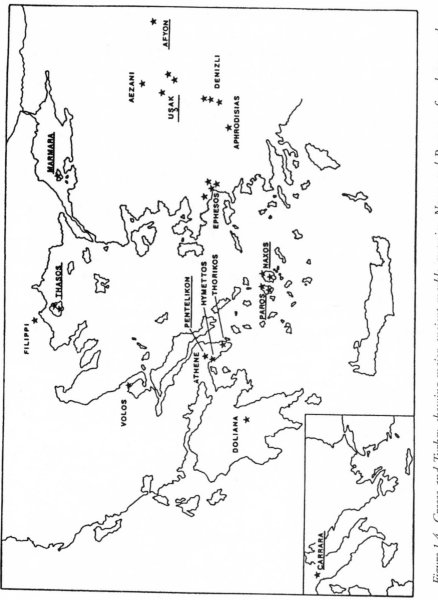

Figure 1.4. Greece and Turkey, showing major ancient marble quarries. Naxos and Paros are found among the Aegean Islands. The inset shows Carrara, in Italy.

Scientific contributions to archaeology often follow an uneven path of evolution. No sooner has a new method been suggested than friends and colleagues enthusiastically point out its flaws and limitations. The original method must then be repaired or abandoned. If repaired, the method probably becomes more complex but more reliable. Eventually an entirely different method may supersede it. The isotopic analysis of classical marble has moved through all these stages except the last one. The original comparison of carbon and oxygen isotopes by the Craigs examined only the four most prominent Greek sources, and their separation in a plot such as Figure 1.5 is simple and elegant. On one axis is the ratio of oxygen-18 to oxygen-16, and on the other, the ratio of carbon-13 to carbon-12. Each of the sites is clearly distinguished, and Naxos even exhibits two distinct quarries. No matter how creative and noble, an idea, when placed in the scientific marketplace, will be subjected to a critical and sometimes bruising review from other scientists (as well as from the originator). After a few years, when numerous other marble sources were studied, the dream of Figure 1.5 became the nightmare of

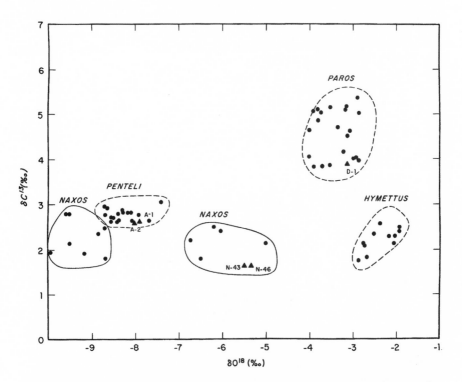

Figure 1.5. A plot of carbon and oxygen isotopes for marble from prominent Greek sources. Each marble source is distinguished in these two dimensions. Reprinted with permission. © 1972 by the American Association for the Advancement of Science.

Figure 1.6. In one spot no fewer than six sources are overlaid and could produce a given isotopic signature. Paros is overlaid by Ephesus. Hymettos and one of the Naxos sources are overlaid by two or three other sources. More recent studies have overlaid the other Naxos source and Mount Pentelikon with other quarries.

Such observations do not have to mean abandoning the method. Rather they are an inspiration to improve it. Just two factors, carbon and oxygen isotope ratios, may be an insufficient basis to identify an unambiguous quarry. Numerous other factors have since been developed, including the isotopes of strontium and the proportions of a variety of elements, including esoteric ones like samarium and scandium, and more familiar ones like chromium and antimony. Particular forms of manganese also proved useful when studied by the electron spin resonance technique (described in Chapter 8). When all these factors are brought together, sources of marble can be specified rather reliably. And what about Antonia? The original analysis of Herz, when reconsidered in the more rigorous context, was found to need no real change.

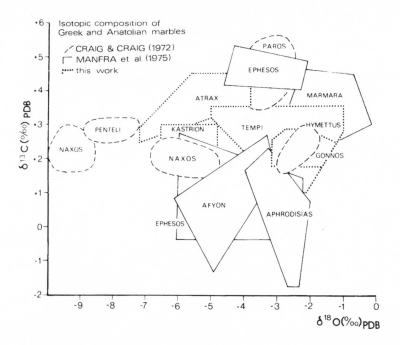

Figure 1.6. A plot of carbon and oxygen isotopes for marble from many different ancient sources. One of the Naxos quarries and the Penteli quarry have no overlap with other quarries. Most of the remaining quarries may be located in Figure 1.4. The new sources offer serious overlap with classic Greek sources—for example, Hymettus is overlaid by Gonnos and Tempi.

Flint

Whereas some of the greatest aesthetic creations of the ancient Mediterranean were made of marble, flint was the most utilitarian stone for earlier Europeans. During the Old Stone Age or Paleolithic Period, humans used flint for a variety of purposes, such as cutting and pounding. At first these uses were served by flint found on or close to the surface. With the advent of farming during Neolithic times, apparently the need for flint tools expanded, and the search for raw materials turned to mines and quarries. Flint generally is found embedded in chalk formations. At accessible locations, serious industry arose. The hatched portions of Figure 1.7 show the widespread locations of chalk

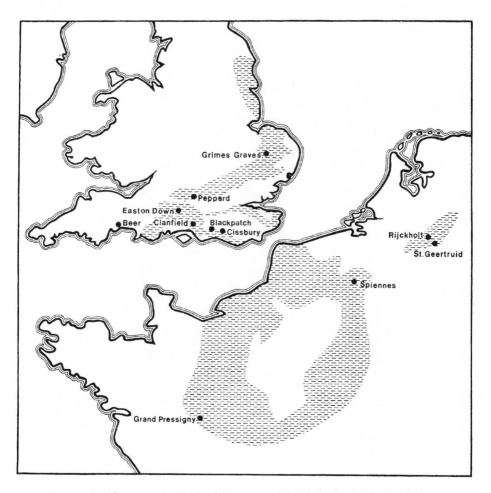

Figure 1.7. Flint mines in England, France, and the Netherlands. The hatched areas represent chalk outcrops.

outcrops in western Europe, particularly France, Great Britain, and the Netherlands. The English sites have such impressive names as Grimes Graves and Blackpatch.

The miners dug a shaft of some 30 to 50 feet, down to the level at which the flint would be found. From the bottom of the shaft, horizontal corridors and galleries were excavated, only 2 feet high but extending maybe 10 to 20 feet away from the shaft. Excavators have found no signs of artificial illumination (probably because it would deplete oxygen), so miners had to make do with whatever indirect light was present. Metal tools were unknown to these people, so they dug with flint axes mounted on wooden handles.

Production at these mines was exceedingly impressive. Figure 1.8 gives an idea of the extent of the English site of Grimes Graves in East Anglia, which was mined from about 3000 to 1500 B.C. Each depression is the remnant of

Figure 1.8. The Neolithic flint-mining area of Grimes Graves in East Anglia, England. Each depression indicates the top of a shaft, with the ancient pilings surrounding it.

the mouth of a shaft. G. de G. Sieveking, who excavated the site, estimated that there were 350 to 500 of these shafts at Grimes Graves, more than 250 at Easton Down, and at least that many at the nearby site of Martin's Clump in Hampshire. If some 30 tons of flint was excavated from each shaft, there was the potential for the production of millions of flint axes. It is no wonder that these axes are so common archaeologically. The figures for the English sites are dwarfed by those for the Dutch site at Rijckholt, which dates to around 3000 B.C. Its excavator, P. W. Bosch, estimated that it contains some 5 thousand mining shafts and produced 15 hundred axes per day for 5 hundred years, for a total of more than 150 million axes. By the end of the Neolithic Period, the demand for polished flint axes had fallen off, and archaeological sites contain a larger variety of tools and weapons made of flint.

Flint, like quartzite, is composed largely of silica (silicon dioxide) and probably was formed by the gradual replacement of carbonaceous (carbon-based) chalk with the chemically more stable siliceous (silicon-based) flint. Because flint occurs in only a few locations but the need for flint by early farmers was universal, mechanisms developed to transport the raw material or the finished product. Understanding this process is where the chemist comes in. Identifying a flint ax as coming from a particular mine or quarry can provide information about early communication and trade routes. The silica of flint is composed of small crystals interspersed with pores that could fill with water. This water could bring in a variety of trace elements that were characteristic of the geological environment. Once again, analysis of the concentrations of these chemical traces can characterize particular mining sites and hence make possible the identification of the provenance of flint axes.

Sieveking and his colleagues suggested this procedure in 1970, and it has held up well over the years. They originally examined the seven British and two French sites shown in Figure 1.7, and later expanded it to more than double that number. They measured the amounts of up to eight trace elements, such as aluminum, iron, and sodium, in flint axes. They constructed an elemental signature for each site, then tested how accurately they could assign specific axes to the sites by examining axes from known sources (found, for example, in the mining shafts). Without any other information, they could successfully identify the source of 70 percent of unknown axes. If other information as well was used, such as archaeological context or date, the figure rose to 95 percent. Such analyses were particularly useful for sites such as the impressive Maiden Castle, which has no local sources of flint. The analysis left fewer than 20 percent unassigned axes from Wessex and southwestern England. In East Anglia the figure is about 33 percent, suggesting that a major source or several minor sources have not been discovered.

Obsidian

Only an archaeologist or a Neolithic farmer might consider a flint ax beautiful, but there is no question that obsidian, yet another form of silica (with other elements), can be beautiful as well as utilitarian. During the European Neolithic Period, obsidian was a material of choice for the manufacture of a variety of cutting tools. As metals replaced obsidian, it still found widespread ornamental use in the production of vases and statues. Phidias, the sculptor of the Parthenon, used obsidian to decorate the statue of Zeus at Olympia (one of the Seven Wonders of the Ancient World). In Mesoamerica, which extends from central Mexico to Central America, the widespread availability of obsidian inspired numerous uses, including extraordinarily beautiful statues, masks, and mirrors, as well as extremely effective knives, which the Aztecs used for human sacrifice at the time of the European Conquest.

Already in the 1960s Colin Renfrew recognized that obsidian was a nearly ideal material for source characterization by elemental analysis. In contrast to flint, obsidian is produced only by volcanic action, so the sources are few and usually rather obvious. A given volcanic flow is very homogeneous, so that the profile of its trace elements can be uniquely characteristic of a given source. Moreover, obsidian is stable to chemical alteration except at its surface, so that analyses of the interior of an obsidian artifact can reliably reflect the composition of the original source flow. As a result, obsidian has been examined chemically to characterize provenance and to define trade routes, possibly more than any other type of stone.

The presence of obsidian far from any sources of volcanic activity provides a particularly intriguing puzzle to the archaeologist. The Midwest and Great Plains of the United States are well known for their unmitigated flatness, certainly not for volcanic activity. Nonetheless, obsidian is commonly found at midwestern pre-Columbian mound sites. As early as 1848, the antiquarians E. G. Squier and E. H. Davis, in their report to the Smithsonian Institution on such sites, recognized the incongruity of such finds and pleaded, "Whence was this singular product obtained?" Over the next 130 years numerous answers to this question were suggested, ranging from Mesoamerica and Peru to the Pacific coast and Alaska. Adon Gordus finally provided a firm answer by carrying out chemical analyses of all these sources, as well as others. Once the sources had been characterized, he analyzed almost a hundred samples from Hopewell mounds in Wisconsin, Illinois, Indiana, and Ohio. The profiles matched those of two sites in Yellowstone National Park, some 1,500 miles west of the finds.

Sometimes obsidian trade routes can be defined in great detail, and in the process some healthy scientific controversies have arisen. The Classic Maya civilization that flourished between about A.D. 250 and 900 was concentrated in the lowland areas of the Yucatán Peninsula (Figure 1.9), but its obsidian came from volcanic flows in the Guatemalan highlands. In particular, the sources in El Chayal and Ixtepeque appeared to dominate. An interesting trade pattern emerged from the chemical characterization of these sources and of artifacts from Classic Maya sites. El Chayal appeared to provide obsidian to the interior sites, such as Seibal and Palenque, via river traffic, such as the Usumacinta might provide. The Ixtepeque source, on the other hand, could send obsidian down the Motagua River to the Caribbean coast and serve sites in Belize and farther along the Yucatán coast. Such a simple model, however, unraveled when some samples from Seibal and Palenque proved to be of Ixtepeque origin, and almost all the samples from the coastal site of Moho Cay were from El Chayal. It is now clear that there was competition between sources, and the sites along the Yucatán coast may have served as connections between them. These directions are indicated by the arrows in Figure 1.9.

European and Southwest Asian obsidian was examined very early by Renfrew for sourcing by trace element content. Sources include the various volcanic islands such as Sardinia and Melos, as well as sites in Turkey, Armenia, Hungary, and Slovakia. African sources in Kenya and Ethiopia have been known since antiquity. (Pliny mentions obsidian from Abyssinia, now Ethiopia.) Renfrew emphasized elements such as zirconium and barium. Later N. H. Gale included not only elements such as strontium (Sr) and rubidium (Rb), but also the isotope ratios of strontium, much as was done with marble. Figure 1.10 shows ellipses based on Sr and Rb for many European and Southwest Asian sources; points on the figure correspond to artifacts from actual dig sites. In a recent synthesis of thirty years of obsidian chemical analysis, O. Williams-Thorpe constructed a map (Figure 1.11) incorporating most known western Mediterranean sources. Arrows radiate from the sources to show trade routes to archaeological sites.

It is worthwhile to take a step back now and ask under what conditions chemical analysis can be expected to provide a successful test for the source of an artifact. First, every source must be found, but that is unlikely. If we do not know all the sources, some artifacts may be left unassigned, as with the East Anglian flints, or they may be assigned incorrectly to the closest known match. Such a conclusion could result in misassigned trade routes. Second, each source must be different from every other source. We have seen in Figure 1.6 that this ideal, however, is not always realized. A third requirement is that a given source must have exactly the same chemical profile for all samples. Unfortunately variability can occur. Fourth, processing by human workers at the time the artifact was made must not have altered its chemical makeup. Rarely is this a problem for stone artifacts, but the processing of metals, which can involve combining and melting down materials from several sources, ren-

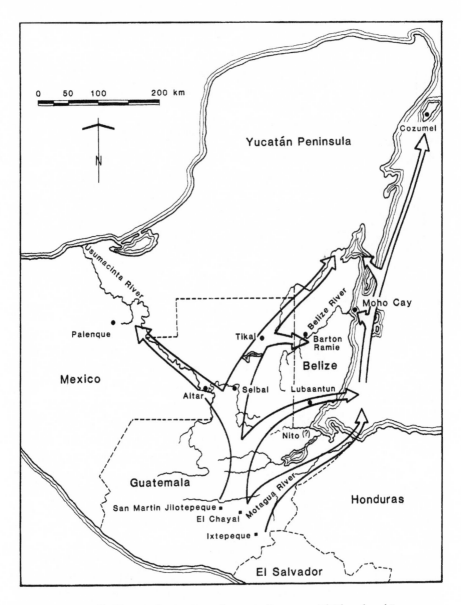

Figure 1.9. The Yucatán Peninsula and surrounding areas. El Chayal and Ixtepeque were primary sources of obsidian for these areas. The arrows indicate possible trade routes as established by chemical analysis of obsidian. Reprinted with permission. © 1984 by the American Association for the Advancement of Science.

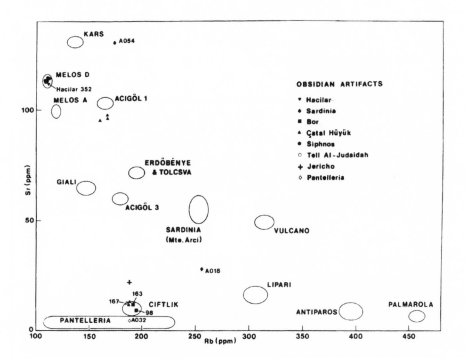

Figure 1.10. Sources of obsidian found in the Mediterranean and Southwest Asia. The axes are the amounts of strontium (Sr) and rubidium (Rb) in the samples. The ellipses represent the range of concentrations of these elements for specific sources, labeled on the diagram. Artifacts are similarly characterized and placed on the diagram, with filled circles representing obsidian from Siphnos, filled squares from Bor, and so on, as given in the legend.

ders them almost impervious to analysis for source in this fashion. Finally, burial conditions must not alter the chemical makeup of the artifact. Again, stone is a sturdy material that is resistant to change even in moist and salty conditions that might be found, for example, on the Yucatán coast. Other materials, such as bone or glass, can undergo alteration while buried, making the chemist work a little harder to provide answers to archaeological questions about trade routes.

Soapstone

Soapstone, or steatite, is a silicate mineral quite different from obsidian. It is pleasantly slippery to the touch (hence the name) and is easily worked because of its softness. It was widely used in the eastern part of the United States from 3000 to 1500 B.C. by what are called Archaic and Early Woodland peoples.

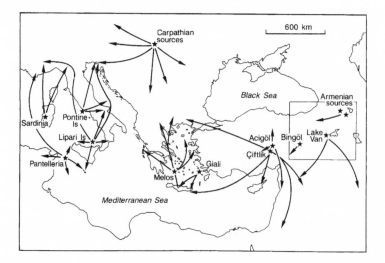

Figure 1.11. Map of the western Mediterranean region, with obsidian source areas represented by stars and names. Arrows emanating from the sources indicate movement of obsidian from the sources to archaeological sites.

Soapstone artifacts are found in habitation sites from these periods—not only utilitarian objects such as bowls, but also decorative objects such as statues. Soapstone, unlike obsidian, is formed through geological reworking or metamorphosis from older rocks, such as serpentine. Its more complex chemical structure is composed of silicates containing a lot of magnesium. The geological transformation can result in characteristic profiles of trace elements.

Soapstone has been studied extensively by Ralph O. Allen, who performed an original twist by basing his conclusions on the proportions of the rare earth elements (REE) present in soapstone. Chemists have not figured out a simple way to display these elements, also called the lanthanides, in the Periodic Table, so they have been relegated to the basement (at the bottom). They include elements with strange names (praseodymium, dysprosium) and others all named after the same town (Ytterby) in Sweden, with a singular loss of originality (ytterbium, terbium, erbium, as well as the non-REE yttrium). (Compare this homogeneity of names with the hot competition for naming new transuranium elements, such as in kurchatovium versus rutherfordium for element 104.) The REE are particularly sensitive to metamorphic processes and hence should be useful in assigning sources for soapstone.

Allen has sampled numerous soapstone quarries in the eastern United States and compared their REE profiles with those of artifacts. In Figure 1.12 the quarries are designated by A through U and are located where the letters are

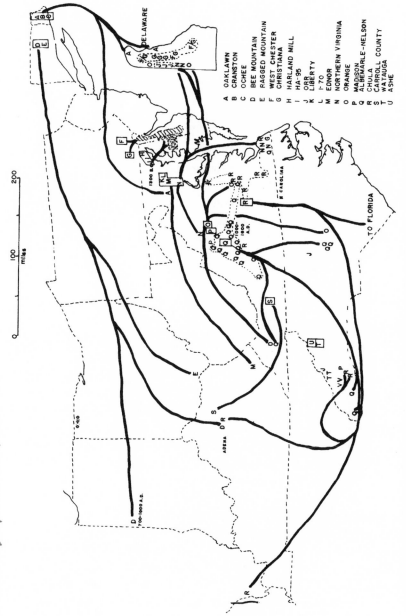

Figure 1.12. *Movement of soapstone from sources primarily in Virginia, North Carolina, and Connecticut. Sources are represented by letters contained in boxes. Archaeological finds that correspond to specific sources have the same letter but lack the box. Reprinted with permission. © 1978 by the American Chemical Society.*

A OAKLAWN
B CRANSTON
C OCHEE
D BEE MOUNTAIN
E RAGGED MOUNTAIN
F WEST CHESTER
G CHRISTIANA
H HARLAND MILL
J HA-95
K ORR
L LIBERTY
M I-70
N EDNOR
O NORTHERN VIRGINIA
P ORANGE
Q MADISON
R ALBEMARLE-NELSON
S CHULA
T CARROLL COUNTY
U WATAUGA
   ASHE

contained within boxes. Unboxed letters identify the source quarries of archaeological soapstone at the location of the find, as based on chemical analysis. The connecting lines then indicate approximate trade routes. Thus sources in Connecticut provided soapstone for sites in Indiana and Kentucky, where the raw material is not found. The map overemphasizes objects traded over long distances. Actually, Allen found that most artifacts were found within 100 miles of the source quarry—in what he called the minimum effort hypothesis. Long-distance trade generally followed a southerly and westerly route. Rarely were artifacts found more than 60 miles to the north or east of the quarry. The enormous quarries in Albemarle and Nelson counties in Virginia (source Q on the map) were the source not only for a large part of Virginia, but also for much of North Carolina. Not shown on the map are other rich sources in Georgia and Alabama, which provided soapstone for much of the Southeast.

Turquoise

Ancient peoples were able to trade over immense distances when particularly valuable goods were concerned. We have already seen how obsidian moved from the volcanic sources in Yellowstone National Park to numerous sites in the American Midwest. Also impressive is the story of turquoise, which has been documented in detail by Garman Harbottle and his archaeological collaborators. Turquoise is a phosphate mineral rather than a carbonate or a silicate. Much of the work on it was inspired by its widespread use in Mesoamerica. When Cortés arrived in Mexico in 1519 and was mistaken for a returning mythical god, he was presented with, among other things, a mosaic turquoise mask. Smaller articles such as rings, beads, and pendants were made out of whole turquoise, whereas larger articles like the mask were constructed as mosaics. Small pieces called tesserae were prepared, beveled, and fitted in such a way that little space was present on all of the irregular sides; and finally glued, usually onto a wooden base. The most puzzling aspect about this enormous industry was the almost complete absence of turquoise mines in ancient Mesoamerica. The mines were located in the desert areas of the American Southwest, from California to New Mexico, and in Chihuahua, on the Mexican side of the border (see Plate 1). Was it possible that these areas, whose cultures appear to have benefited so little from those of Mesoamerica, provided the enormous bulk of turquoise crafted in Mesoamerica?

To answer this question, Harbottle analyzed a dozen or more samples from some forty mines in the Southwest and compared them with the elemental profiles of some two thousand archaeological samples from twenty-eight sites, both in the Southwest and in Mesoamerica. The results provided the trade routes that are drawn out on Plate 1. Indeed, southwestern turquoise penetrated not only to central Mexico, but even to the Yucatán Peninsula. During the Classical

Period, with a peak around A.D. 700, the Mexican site of Alta Vista was a thriving workshop for turquoise, transforming the raw material primarily from the mines at Cerrillos in New Mexico. Turquoise from Cerrillos has been positively identified in the Mexican states of Sinaloa, Jalisco, and Nayarit.

Turquoise began to be worked closer to its southwestern sources during the Pueblo II Period, around A.D. 1100, particularly at Chaco Canyon. By the mid-thirteenth century, numerous new mines had opened, making possible the massive use of turquoise not only in the Southwest, but also by the Aztecs. Harbottle has documented both a Pacific coastal trade route and a more inland route.

Basalt and Sandstone: Stonehenge

Utilitarian basalt and sandstone represent the opposite extreme from decorative turquoise and obsidian. They were used in particular for large structural units such as the standing stones of numerous prehistoric sites in western Europe. At this juncture, in order to consider these materials, a little information about geology will be useful. Many types of rocks, called igneous, are formed by volcanic action, including obsidian and basalt. Others, called sedimentary, are formed by the accretion of marine sediments. Thus quartz sand becomes sandstone, and calcium carbonate, from shells and coral, becomes limestone. These materials may be reworked over geological time through the effects of pressure, heat, and water into new materials, called metamorphic rock. Thus soft sandstone and limestone are metamorphosed into much harder quartzite and marble.

This background information helps to appreciate the complex problems associated with understanding the construction of the magnificent ruins of Stonehenge, on the Salisbury Plain in southwestern England. The various stages of construction lasted about a millennium, and the remains that we see today were in place by about 1800 B.C. The dominating outer circle and enclosed horseshoe are composed of giant sarsen stones with crosspieces (lintels), which are made of a sedimentary quartz or silicate. The smaller stones of the inner circle and the innermost horseshoe have been called bluestones and are largely basaltic.

As early as 1720 Edmond Halley, of comet fame, examined these stones and concluded from their general wear that they had been 2,000 to 3,000 years in the field. In 1923 H. H. Thomas examined the basaltic dolerite bluestones and concluded that they could have come only from the Preseli Hills of southern Wales, some 240 kilometers to the east. The Salisbury Plain is devoid of such stones, and Thomas felt it was unlikely that they could have been transported by glacial action. The enormous effort required to transport the stones over such a distance led Colin Renfrew to suggest that up to 30 million man-hours were required to assemble the entire site.

Thomas used standard methods of petrology (the science of rocks), such as microscopic examination, to demonstrate the identity of the Stonehenge and Preseli materials. There is no doubt that they are indeed the same, but O. Williams-Thorpe and R. S. Thorpe now have presented a revisionist view, also based on sound science, but tempered with common sense and good archaeology. They focused on the bluestones: thirty of them are basaltic dolerite, five are igneous rhyolite, five are other volcanic materials, and three are sandstone. Instead of simple microscopic petrological comparisons, they carried out extensive chemical analyses, both of the bluestones and of possible geological sources, to uncover elemental fingerprints. Titanium and zirconium, for example, were particularly useful in making source distinctions. They found that the eleven dolerites they analyzed came from at least three different sources in the Preseli Hills and that the rhyolites came from four other nearby sources, up to 30 kilometers apart. Thus, if the Stonehenge builders had hauled their materials from eastern Wales, they had not chosen a select quarry, but indiscriminately used numerous sources.

When Williams-Thorpe and Thorpe reviewed the geological record, they concluded that the Salisbury Plain was involved in glaciation, both from the north and from the east. The Bristol Channel Ice in fact would have followed the precise route needed to move Preseli materials to the vicinity of Stonehenge. They also reviewed the archaeological literature of British and Continental prehistoric megalithic structures and found that stones rarely were transported more than 5 kilometers by the builders. Given that the Anglian glaciation of some 400,000 years ago could have transported the bluestones to the Salisbury Plain, Williams-Thorpe and Thorpe concluded that the builders of Stonehenge used nearby glacial erratics (randomly dropped stones). Although the giant sarsens have never been analyzed chemically, they are widely available in southern England, and also probably were available locally.

The proof is not perfect, but the arguments of Williams-Thorpe and Thorpe are persuasive. The new interpretation, however, detracts little from the accomplishments of these early British inhabitants. They still had to move the stones several kilometers and erect them in an aesthetically pleasing and astronomically useful fashion that astonishes visitors even today, in its partially ruined state.

Basalt has been used throughout history to produce objects of great size and weight. Although the bluestones of Stonehenge are one example, another involves basalt that had been formed into large millstones for grinding grain. Such objects were found off the small island of Sec in Palma Bay, off the island of Mallorca. Excavation indicated that they came from the wreck of a fourth-century B.C. Greek ship. Chemical analysis combined with archaeology has reconstructed the route taken by this ship before it sank. Its excavated cargo contained a large number of amphorae (pointy-bottomed vessels used to transport and store oils), bronze vessels, and millstones. While the archaeologists characterized the amphorae stylistically, the same team of geochemists who worked on Stonehenge, O. Williams-Thorpe and R. S. Thorpe, carried out the elemental analysis of the basaltic millstones. The result of this collaboration was an impressively detailed re-creation of the route of the Sec ship during its last voyage (Figure 1.13), made possible because Williams-Thorpe and Thorpe had access to sources of basalt that had been previously characterized (by elemental content) all around the Mediterranean.

As the ship was of Greek style and the cargo included 150 Samian amphorae, it may have begun its voyage at the Greek island of Samos. One of the millstones had the elemental pattern of the nearby island of Nisyros. The cargo contained several dozen millstones of the type called hopper-rubbers, whose elemental profile indicated they had come from the island of Pantelleria, located between Sicily and the African coast, near the ancient city of Carthage. The authors suggested that the ship offloaded much of its cargo in Sicily and took on the Pantellerian millstones either as ballast or for future trade. Alternatively, a stop could have been made at Pantelleria itself. The bronze objects may have been taken on at Carthage. One of the millstones of a different type had an elemental profile from Mulargia in Sardinia, and also could have been taken on at Carthage. It is puzzling that the ship then moved west rather than east back to Greece. Inclement weather might have blown it well off its course and resulted in the final wreck, but among the amphorae were some from the Mallorcan island of Ibiza, suggesting the ship had made at least one stop already in the vicinity where it sank. The archaeological study of the amphorae and the chemical study of the millstones thus suggested the ports of call of this ancient Greek trading vessel.

Limestone and Calcite

Next to silicate materials like basalt, obsidian, and flint, the most common minerals in the earth's crust are carbonates like marble and chalk. Limestone is a form of calcium carbonate that has been used for millennia in building blocks for walls. In ancient Greece softer limestone preceded marble as the material for sculptures. In 1926 the Cleveland Museum of Art purchased a Late Archaic Greek limestone sculpture said by the dealer to have come from the Athenian Acropolis. Such a loose attribution can hardly be considered solid provenance, but it led art historians and chemists on a chase that eventually ended in a blind alley.

The sculpture is in the shape of a goat's head (Figure 1.14), and therefore could have represented the Arcadian herdsman's god Pan, who was often

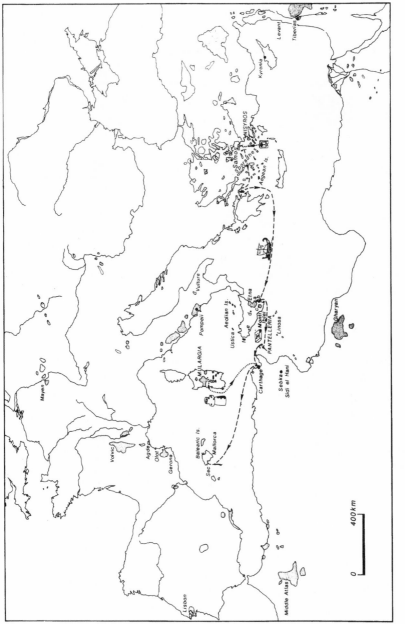

Figure 1.13. The Mediterranean Sea, showing the hypothesized route of the Sec shipwreck. Names entirely in capital letters are basalt sources proposed for the millstones found in the wreck. The ship may have begun at Samos, stopped in mainland Greece, Sicily, and Carthage, and finally foundered near Mallorca.

Figure 1.14. Limestone head of Pan, from the Cleveland Museum of Art.

depicted as part-man, part-goat. It was then an easy step to speculate that the head might have come from a no-longer-existing monument on the Acropolis attributed to Miltiades. The story of this monument was given by Herodotus. In 490 B.C., as the Persian fleet threatened Athens, the famous Athenian long-distance runner Philippides was sent to Sparta to ask for help. On the way Philippides was said to have been accosted by Pan, who offered to help Athens if they showed more interest in him. After the victory over the Persians at Marathon, the Athenian leader Miltiades had a monument built to Pan for his alleged help. The only two authentic fragments of this monument, now named after Miltiades, reside in the Acropolis Museum in Athens and have no great aesthetic appeal. If the Cleveland sculpture were from the Miltiades monument, it would be not only the best preserved extant piece, but also a unique artistic residue of the early wars between the west, represented by Athens, and the east, represented by Persia. Enter the chemist.

The Pan head and the two Miltiades fragments in Athens were similar in color and texture, but the elemental profiles were disappointingly different. The amounts of arsenic, chromium, and uranium, for example, were about the same in the two authentic Miltiades fragments, but five times that in the Pan. Samples from limestone sources in Attica, near Athens, also were examined and found to resemble in elemental profile the two fragments, but not the Pan. Thus the fundamental question was answered: the Pan had not come from the Miltiades monument. Its origin is still a mystery.

Limestone was used extensively by medieval craftsmen in the construction of cathedrals across Europe. Garman Harbottle and his colleagues, the same group that examined turquoise, have compiled a large database on limestone with emphasis on medieval European structures. For example, they were able to pinpoint that the stone used to build and decorate the Nôtre Dame Cathedral in Paris came from the nearby quarry at Charenton and not from various other quarries in the vicinity, such as Conflans-Sainte-Honorine or Carrières-sur-Seine, all within 15 miles of the center of Paris. M. A. Bello and A. Martín had a more daunting task in identifying the source of the building stone for the Cathedral of Seville. Not far from Seville are numerous quarries that could have supplied the stone (Figure 1.15), which is a mixture of sandstone and limestone. (There is more silicon in sandstone and more calcium in limestone, so the former is called siliceous and the latter calcareous.) Although there is no specific documentation of quarry sources, the cathedral archives imply that more than twenty quarries could have been involved in its construction. The wide variation of the ratio of silicon to calcium oxides, from nearly pure limestone to highly siliceous material, substantiates the multiplicity of sources. The most commonly used materials, with 20–45 percent of both silicon and calcium oxides, were found by the chemists to come from the single quarry at Puerto de Santa María, near the coastal city of Cádiz. The ease of transporting it along the coast and up the Guadalquivir River, together with its appropriate strength, probably made this material the primary choice.

Limestone is also commonly found in caves. Because calcium carbonate is acid soluble, acidic water that percolates through limestone deposits can dissolve them and create caves. Water containing calcium carbonate drips down from cave ceilings, evaporates, and leaves a buildup that we call stalagmites and stalactites. A strategically placed skull in the Petralona Cave in Greece was cemented to the floor by dripping water and was encrusted with a form of calcium carbonate called calcite. The skull's age was of interest because it showed characteristics that are intermediate between the early human species known as *Homo erectus* (predecessor of the modern *Homo sapiens*) and the famous

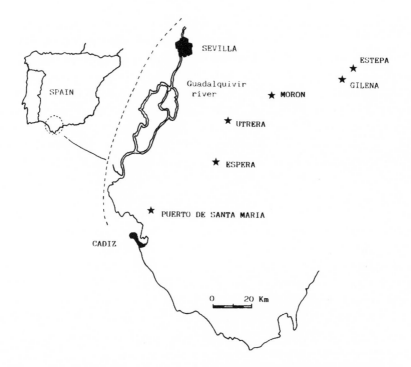

Figure 1.15. That portion of Spain containing Seville, Cádiz, the Guadalquivir River, and several possible quarry sites for the stone used to build the facade of the Seville Cathedral.

Neanderthal or Neandertal humans (considered by many to be an evolutionary dead end). Any information about the transitions between these forms, such as the skull seemed to offer, would be of vital interest. The key scientific input would be the date of the skull. There are several reliable methods of dating calcite. Chemistry came in by showing the relationship between the specific dated layers of calcite and the calcite covering the skull, which was harder to date. The pattern of trace elements identified the calcite on the skull with larger amounts of calcite in the vicinity on which the dating procedures were carried out. The current estimate is that the skull is at least 160,000 to 200,000 years old. An upper limit is still controversial, but it may be 350,000 years old or more. These dates bracket the period when *Homo erectus* began to evolve into other species related to or leading to modern humans.

Carbon-14 Dating: Rock Varnish

Up to this point, the elemental content of stone artifacts has been used as a signature of the raw materials from which they were made. It is assumed that the stones did not change over time. Although this is a reasonable assumption for the interior or bulk of the stone, surfaces do change. Surface material, therefore, is usually discarded before chemical analysis is carried out. In doing so, are we throwing out valuable information? Can the surface changes also be useful? Geologists have studied surface structure extensively, but only recently has such information begun to be useful archaeologically. When a piece of flint is chipped to form an ax or blade, pristine interior surfaces are exposed. Any changes in this new surface might provide useful information.

When a stone is exposed to air, color changes occur slowly (Plate 2). The colored material that builds up is called desert or rock varnish, or simply patina. Although there may be a variety of processes that result in these surface changes, Ronald I. Dorn has demonstrated that often it is a surprising interaction between the inorganic and biological worlds. Colors include bright orange, which Dorn's chemical analysis showed comes from the presence of iron oxides; and browns and black, from the presence of manganese oxides. The puzzle was how these specific elements could be concentrated in levels far above those present in the surroundings. A complete analysis of rock varnish by Dorn revealed that it is composed of about 60 percent clay (aluminosilicates), 20–30 percent iron and manganese, and the remainder, a variety of other metal oxides. Because rock varnish appears to form more quickly where rock is moist, a biological source was suggested. Examination of the varnish from this point of view revealed the presence of bacteria such as *Metallogenium* and *Arthrobacter;* the former is known to concentrate the element manganese.

The following scenario has been suggested by Dorn. Encouraged by moisture and stabilized by adhering clay, bacteria begin to colonize the newly formed surface of the stone. In harsh environments bacteria that rely on organic nutrients do poorly, but bacteria called mixotrophs, which derive some of their sustenance by oxidizing manganese to its oxides, can make do. The result is slow bacterial growth on the stone surface that eventually imparts new colors to it. These changes may be exploited by the archaeologist to determine how much time has elapsed since a surface was clean—when a flint or obsidian was chipped, or when a petroglyph was carved into the surface.

The presence of bacteria also provides a source of carbon for the carbon-14 dating technique. Although well over 99.99 percent of carbon is made up of the stable isotopes called carbon-12 and carbon-13, a small amount (carbon-14) is unstable or radioactive. Carbon-14 is formed in the upper atmosphere by high-energy cosmic rays, and the relative proportions of stable and radioactive carbon are relatively constant throughout the world. A living organism uses atmospheric carbon as the building block of all organic molecules, so that radioactivity is naturally incorporated during life at very low levels. At death this incorporation ceases. As the radioactive nuclei decay, the amount of carbon-14 decreases. The levels of carbon-14 may be measured either by radioactivity detectors or by mass spectrometers. These levels may be converted to actual dates for the time when the organism died. The mass spectral method requires less than a milligram of material.

Carbon-14 levels for the bacterial residues of rock varnish closest to the surface of the rock provide a measure of the time at which the surface was exposed. In this way Australian petroglyphs have been demonstrated to be at least 36,000 years old. The Nazca lines in Peru (Figure 1.16) also have been dated in this fashion. Very large geometric figures, many in the shape of animals, were formed in the Pampa San José in southern Peru by removing varnished rocks from the surface and exposing unvarnished underlying rocks. The subsequently built-up varnish on the exposed rocks was dated, by the carbon-14 method, at 190 B.C. to A.D. 660, a range that corresponds to the Nazca Culture, which had already been well studied by archaeologists. These dates put to rest any idea that the lines had been created by a much more ancient culture. The lines, nonetheless, are impressive aesthetic monuments to the Nazca Culture and may have been constructed to honor gods who were

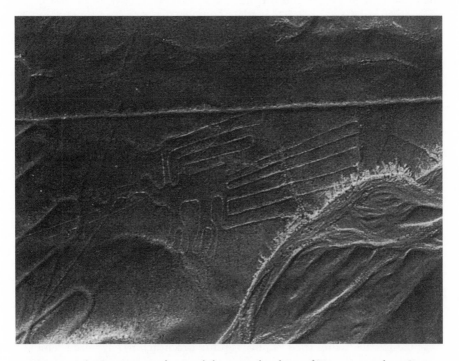

Figure 1.16. Geometric and animal shapes in the plains of Nazca in southern Peru. The lines were constructed by removing varnished rocks from the desert floor and exposing unvarnished rocks.

considered to reside in the sky, just as western European cultures have constructed cathedrals with tall steeples.

Stone Monuments, Air Pollution, and Conservation

Trace elements have been very successful in uncovering the sources of many types of stone, but an important assumption has been that the stone did not change between its initial working and the chemist's analysis. Stone is subject to chemical change, and the problem increases for stone exposed to a polluted atmosphere. Building stones, for the most part, are either calcareous, such as limestone and marble, or siliceous, such as basalt, sandstone, and granite. We have already discussed the chemical analysis of all these materials but granite, whose highly heterogeneous nature has not yet lent itself to this type of analysis. All building stone is susceptible to the vicissitudes of physical destruction by war, urban renewal, earthquake, or strangulation by tree roots. Chemical analysis is unfazed by such events, but chemical destruction is another matter. Chemical change tends to occur from the surface down, so that most stone materials remain unchanged on the inside, even when sitting on the ocean floor for several thousand years. Analysis of the interior of a stone thus provides a reliable fingerprint. Surface alteration, however, can have profound effects on both the aesthetic and structural properties of stone.

Modern air pollution has aggravated these problems, which have become increasingly obvious, whether in the highly publicized decay of the marble of the Acropolis or the Taj Mahal, or in the more personal decay of the tombstone of one's great-grandparents, on which irreplaceable family information had been recorded. Stone made out of calcium carbonate is the most susceptible to this type of decay, because the basic carbonate may be neutralized by acid from rain or the soil and form soluble salts that are washed away. In the most common such process, sulfur in coal is burned to form sulfur oxides, which adhere to stone surfaces and are transformed, either chemically or with the help of sulfur-eating bacteria, to the sulfate form. The calcium sulfate ($CaSO_4$) so formed is more soluble in water than calcium carbonate. Calcium sulfate as a mineral is called gypsum and is found naturally in great deposits, including the enormous dunes in the White Sands National Monument in New Mexico. Gypsum may appear as a flower of crystals on the stone surface, or it may be washed away, contributing to loss of surface detail. Dissolution of the marble on the Acropolis has reached the stage that the caryatids, the monumental female figures used as pillars in the Erechtheum, have been removed for safekeeping and replaced with reproductions.

Although efforts to control air pollution certainly can slow down these problems, it also is necessary to deal with them directly by cleaning and consolidating the stone surfaces. Pollution on the surface of stone can be an impressive mixture of algae, gypsum, reformed calcium carbonate (as the mineral calcite), and adventitious particulates such as ash and soot. Cleaning consists of brushing with steam and hot water, which naturally removes part of the surface. Treatment with harsh chemical agents such as dilute hydrofluoric acid may be effective, but almost always has drawbacks (effects on the person doing the washing, pollution as runoff, and unforeseen effects on the stone). Chemicals have proved more useful in consolidating stone. Barium hydroxide may be injected into carbonate stone to replace the calcium part of gypsum, since barium salts in general are less soluble and more stable than corresponding calcium salts.

Although carbonate stones are most susceptible to decay, silicate stones are not immune, particularly sandstone, which can have limestone veins (as in the Seville Cathedral) and is particularly porous, with the result that pollutants have easier access. The public antiquities of Cologne, Milan, and Bologna, for example, are constructed largely of porous sandstone, which is subject to chemical decay. For these silicates, the stone structure may be rebuilt chemically. Injection of organosilicates or silicones can help rebuild and seal the silica structure. These chemicals are effective only with silicate-based stones, not with the carbonate-based limestones. These consolidants are derived from the plastics revolution. Small molecules (monomers) react with many other identical small molecules to form a variety of large molecules (polymers). These polymers sometimes are called plastics but include everything from the polystyrene in cups and the fabric in carpets to the synthetic rubber in tires and the artificial ivory in pool balls. For the consolidation of sandstone, specific monomers are injected into the stone and then are made to react with each other to form up into the large polymer, which provides both stability and water-repellent properties.

Such chemical wizardry may seem an ideal cure for the effects of pollution, but the long-term stability of such materials is not known. What will the polymers be like in ten or five hundred years? The conservator must abide by the dictum of Hippocrates: First and above all do no harm. Examples of well-intentioned but tragic conservation efforts are known. From 1902 to 1909 the Acropolis underwent an extensive restoration. To stabilize large stone blocks that threatened to become detached, iron rods were inserted through the heart of the blocks. Although oxidation of iron in the rods to its oxides was not unanticipated, it was not known that the oxidation products of iron, such as limonite, are less dense than iron itself and hence take up more space. As more space was inevitably demanded by the new chemical form, the cracking and decay of the marble was aggravated far beyond the situation present in 1902. With such cruelly taught humility, the conservator and chemist must proceed with extreme care.

Synthetic Stone

The earliest human technologies involved only physical or mechanical alterations of raw materials. Stones were chipped; wood and bone were carved. The finished product had an unmistakable relationship to the raw material. About 14,000 years ago, humans for the first time chemically created a material not found in nature: plaster. From the site of Lagama North VIII in the Sinai Peninsula, small stone tools that had been attached, or hafted, to handles by means of plaster have been dated to this period. The Natufian Culture of the Levant, at the eastern end of the Mediterranean (10,300–8500 B.C.), began to use plaster architecturally, as a floor or wall covering. By the Pre-Pottery Neolithic B Period (7500–6000 B.C.) in Southwest Asia, the architectural use of plaster expanded, and nonarchitectural uses—such as for vessels, statuary, or ornamentation—presaged the manufacture of pottery.

These early Southwest Asian inhabitants thus had learned how to alter the chemical forms of nature. They began with a raw material, either limestone ($CaCO_3$) or gypsum ($CaSO_4 \cdot 2H_2O$), and transformed it into a distinct chemical material, plaster, by heating. Limestone loses carbon dioxide to form lime or quicklime ($CaCO_3 \rightarrow CaO + CO_2$). Gypsum loses most of the water molecules buried in the solid to form what we call plaster of paris ($CaSO_4 \cdot 2H_2O \rightarrow CaSO_4 \cdot \frac{1}{2}H_2O + \frac{3}{2}H_2O$). Limestone requires a temperature of 800–900°C for this change to occur, whereas gypsum requires only 150–200°C. Thus the production of lime plaster is more energetically demanding than that of gypsum plaster. The powdery plaster of paris requires only the addition of water to create a pasty mix that can be applied to a surface. The molecular structure reverts to the hard gypsum form after excess water evaporates. Lime requires more processing. First water is added to form slaked lime or calcium hydroxide ($CaO + H_2O \rightarrow Ca(OH)_2$), which fulfills the same role as plaster of paris. When the pasty slaked lime was used as a plaster (on a surface) or as a mortar (between two surfaces), it reabsorbed carbon dioxide and hardened back into calcium carbonate ($Ca(OH)_2 + CO_2 \rightarrow CaCO_3 + H_2O$). Thus these people had discovered how to convert stone into a more workable material that would spontaneously return to stone under controlled conditions.

Lime plaster was used during prepottery periods in the eastern Mediterranean valleys of the Levant and Anatolia, whereas gypsum plaster was used in the drainage areas of the Tigris and Euphrates rivers (Figure 1.17). Because gypsum plaster is about 0.2 percent soluble in water (lime plaster is only 0.0015 percent), it was practical architecturally only in dry climates. On the other hand, lime plaster was much more costly in terms of energy requirements, which came down to the availability of timber to stoke the fires for converting limestone to lime or gypsum to plaster of paris. Overexploitation of timber resources may have contributed to settlement abandonments in the Levant after 6500 B.C.

Improvements were made in lime plaster over the centuries. Inert fillers such as sand or ground limestone were added to the slaked lime (or binder) to extend the amount of material and to improve its properties as a mortar. Hydraulic lime mortars, which could set under water, were created by adding a third component containing silicates. Additions included clay, volcanic ash, or powdered brick. During the hardening process, lime reacted with the hydraulic component to form silicates or aluminosilicates (clay contains a lot of aluminum). The resulting plaster or mortar possessed superior strength and hydraulic properties and became known as cement. Roman cement was composed of freshly burned lime and volcanic ash. When the filler included pebbles, it resembled modern concrete. The dome of the Roman Pantheon (A.D. 110) is composed of Roman cement and still stands after almost two thousand years. The technique of cement manufacture was lost after the Roman period, until

Figure 1.17. Locations where lime and gypsum plaster were used by the Pre-Pottery people of Southwest Asia. Reproduced with permission of the Journal of Field Archaeology *and the trustees of Boston University. All rights reserved.*

the eighteenth century. Its rediscovery in England led to a material that resembled Portland stone and so was called Portland cement. In the modern manufacture of cement, a mixture of limestone, clay, and other components is heated in a long rotary kiln to 1500–1600°C to form silicates (like Ca_3SiO_5), aluminates (like $Ca_3Al_2O_6$), and aluminosilicates (like gehlenite, $Ca_2Al_2SiO_7$). Curing of cement (the process of drying and hardening) involves hydration of the silicates rather than carbonation of slaked lime, as occurred in the original lime plaster.

Chemical examination of archaeological plaster, mortar, or cement involves both elemental analysis and microscopy. Metals such as iron, magnesium, and strontium, which substitute for calcium, vary from place to place and can be used as a fingerprint for the raw material source. The relative proportions of binder, filler, and hydraulic, as determined by chemistry and microscopy, can indicate level of technology as well as geographical sources.

The cases of mortar and cement illustrate both how chemical technology developed and how we analyze the products of ancient technology today. Early people discovered that they could change nature—first probably to reproduce, in this case, natural stone, but eventually to replicate materials with previously unknown composition and properties. Plaster could be spread over a surface in a way not possible with stone, and its smooth surface was difficult to replicate with natural stone. The binding properties of mortar enabled small stones to be transformed conveniently into large buildings. Finally, cement could be molded into shapes that were both utilitarian and aesthetic. Modern chemical analysis of these materials helps us understand how the technology of synthetic stone evolved. The early technology of these materials set the stage for the development of pottery, glass, and metals, which are more complex technologies. Later chapters consider both how chemistry was involved in their development, and how modern chemistry serves to understand the ancient technologies and to preserve the artifacts.

2

S O I L

Soil in a sense, is the medium for the art of the archaeologist. It contains and conceals, but also constitutes the object of the archaeologist's search. Soil is not simply to be removed, but also to be studied and understood. The archaeologist examines its color, consistency, texture, and stoniness, and microscopically measures the size and shape of its particles. Evidence is weighed for erosion, human disturbance, salt accumulation, waterlogging, and even creep (natural movement). Its overall structure must be assessed in terms of technically defined layers, or horizons. Distinctions are made between soil and sediment. Whereas soil has a large organic component from decayed organisms and is found in place, sediment is primarily inorganic and is transported by water or wind from elsewhere.

These measurements are carried out visually and physically, usually without the assistance of the chemist. The presence of human activity, however, has profound effects on the chemical makeup of soil. Chemical analysis of soil or sediment can assist in finding sites or features within sites, in determining what activities took place at a site, in analyzing what happened to material while it was buried, and in locating the source of raw materials from which artifacts were constructed.

Prospecting with Phosphorus

Although the term *prospection* may first call to mind gold rushes, with images of California, the Klondike, and Australia, archaeologists also prospect. To help focus their excavation, they employ an array of techniques. These include walking the site, measuring the electrical resistance or magnetic properties of soil (which are sensitive to human-induced changes), taking aerial photographs, and bouncing radar below ground or below the surface of a body of water to reveal large objects. To this array the chemist adds elemental and organic analysis. Human habitation adds certain chemical elements to natural soil, including carbon, nitrogen, sulfur, and calcium, which, however, tend to return to the natural baseline as time passes. On the other hand, phosphorus in the form of phosphate can be quite resistant to change. In particular, phosphate forms highly insoluble compounds with iron, aluminum, and manganese, which are widely present in soils, particularly clays. These elements help to lock up the phosphate for considerable periods of time, providing a useful chemical residue of human activity that the archaeologist can exploit in searching for and understanding sites of human habitation.

Phosphorus, an essential component of biological molecules, is found in both plants and animals. It is a part of the nucleic acids that make up our genetic material and a part of the biological molecules that store and release energy, as in adenosine triphosphate (ATP). Humans return biological phosphate to the soil in the forms of excrement, food wastes, general domestic refuse, and plant and animal remains (including human burials). Dung, as a working material rather than as a waste product, has been used as tempering in clay floors and as filler in walls made of stone or wattle (poles interwoven with small branches). In addition, nonarchaeological sources of phosphate include modern fertilizers, detergents, and pesticides.

Thus human habitation in general increases levels of phosphate in soil, and the chemical experiment is to survey for phosphate over suspected areas. High levels of phosphate, once identified, then point to locations for further exploration. A grid is set up over the site. Every few meters a sample is taken below the surface (to avoid modern contamination) and possibly much deeper, say every 20 centimeters, down to a depth of a meter. The interpretation of phosphate concentrations is unfortunately complex. Some phosphorus sources, such as bone, are entirely inorganic—that is, the phosphorus is not part of carbon-based (or organic) molecules. Other sources, such as excrement or food waste, are largely organic. Although the method of chemical analysis depends on whether the phosphorus is inorganic or organic, that distinction is not important here. Another source of complexity is that some human activities can even reduce levels of phosphate. For example, overgrazing by domestic animals such as sheep or cattle can lead to erosion and the subsequent removal or leaching of phosphate minerals. Phosphate surveys are most useful when other procedures fail. For example, extensive foliage or rock formations may obscure or prohibit aerial photography. Plowing or other human activities can eliminate almost all visible signs of prior habitation, so that electrical or magnetic surveys are difficult. Even when phosphate analysis is used, however, normally it is not the sole technique for prospection.

Phosphate surveys in agriculture preceded those in archaeology. A Swedish agronomist named Olaf Arrhenius, working for a sugar beet company, first carried out systematic phosphate surveys. Having observed increased levels of phosphate where humans had settled, he suggested that the method could be used to locate lost settlements. For example, the village of Stokkerupp in Denmark was deliberately destroyed in 1670 and its location soon lost. During the 1930s W. Christensen sought to relocate the lost village through phosphate analysis. The approximate location already was known, and he analyzed for phosphate along a series of lines across the site, or transects (Figure 2.1). When the main area of high levels had been located, the extent of the village was circumscribed with a grid of samples. This project involved about 250 samples spaced about 50 meters apart. With today's technology for rapid on-site phosphate analysis, the work would take only a few days, resulting in an excellent indication of where to excavate most productively.

By its scope, the massive flint-mining site of Grimes Graves (described in Chapter 1) indicates activity for more than a millennium, but archaeologists have found no visible remains of an occupation site for the miners (Figure 1.8). The site itself covers about 200 hectares (a hectare contains 10,000 square meters, or about 2½ acres). Forestation has prevented aerial surveys. Sampling for phosphate analysis by a group led by Michael J. Hughes and Paul D. Craddock took three seasons totaling 144 man-weeks, with samples taken with coring rods to a depth of a couple of meters. No major settlement was ever

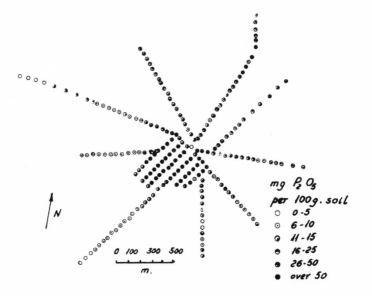

Figure 2.1. Phosphate survey of the village of Stokkerupp, north of Copenhagen, Denmark. Reprinted with permission of John Wiley & Sons, Inc. © 1980 by John Wiley & Sons, Inc.

discovered. This disappointing result after so much hard labor can be viewed positively in two ways. First, the archaeologists could now hypothesize that the site never involved a major settlement. Possibly it was worked by only a few people at a time. Second, the chemists gained considerable experience both in field work and in laboratory analysis that simplified future surveys. They learned how to carry out the analyses quickly on site, so that the archaeological director could help guide the survey and (when results were positive) alter the excavation design. They also learned that deep coring is not always necessary. Unless there is evidence for very long-term and intensive habitation, the soil relatively close to the surface provides sufficient indication of human activity.

This same team of chemists used their experience from Grimes Graves with much better results at Fengate, which is located east of Peterborough on the western edge of the East Anglian fens. A previous archaeological survey had suggested the presence of a farmstead from the Roman period. The area was sampled at a constant depth of 0.5 meters (no deep coring!), on a 10-meter grid. Surprisingly, the phosphate was not particularly high in the Romano-British farm site, but areas at the northern edge of the site, not known to be of archaeological interest, proved to have significant phosphate enhancement. Subsequent excavation revealed an extensive Iron Age (pre-Roman) settlement that had been invisible to an aerial survey because of flood sedimentation. The focus of the excavation then shifted from the Roman-era farm, which actually proved to be rather barren of finds, to the more fruitful pre-Roman site, which was given the name Cat's Water.

These three studies illustrate the full range of possible results of a phosphate survey: successfully locating an expected settlement, failing to locate any settlement, and locating an unexpected settlement.

Vertical Surveys

Coring is carried out in the survey of a large area in order to examine soil materials below surface disturbances or contaminations. When an extensive occupation site is located, the archaeologist may uncover several periods of habitation. Each period has its own layer, or horizon, that is characterized by the appearance of the soil, its chemical composition, and the artifacts or remains present. Because a site may not have been occupied continuously, archaeological horizons may be separated by sterile zones. Phosphate analysis can help sort out sterile and occupied zones through vertical rather than horizontal sampling. Figure 2.2 shows a vertical section for phosphate analysis at Runnymede Bridge, in Egham, England, by the same team of Craddock and Hughes, now also including Michael R. Cowell. Two clear settlement horizons are seen, one from the Late Bronze Age and the other from the Neolithic Period.

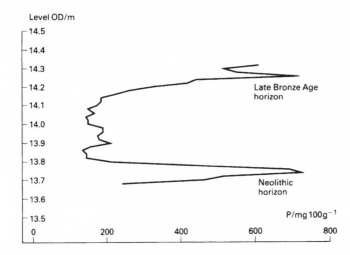

Figure 2.2. Phosphate concentrations in the soil at Runnymede Bridge, Egham, England, shown on the horizontal axis. The vertical axis shows the depth of the point sampled. Two phosphorus peaks occur that correspond to Late Bronze Age and Neolithic horizons. Reprinted with permission.

A. Sánchez, M. L. Cañabate, and R. Lizcano used phosphorus analysis to examine the vertical structure of the Late Neolithic or Early Copper site of Polideportivo in Martos, province of Jaén, Spain, dating to the period between the fourth and third millennia B.C. They selected two occupational complexes that were representative of the major structural phases of the site. The upper part of Figure 2.3 illustrates a cross section of Structural Complex 17. The horizons arose both from activities of the humans who occupied the site and from natural and human-made destructive processes such as rain, fire, and desertion. The lower part of the figure shows the levels of phosphorus sampled at several points in the structure. Background from natural strata represented about 60 milligrams (mg) of phosphorus (P) per 100 grams (g) of material. Chemical analysis showed two major phases of occupation, the horizons labeled IIIA/B and I. Lower phosphorus levels in region IIA possibly indicated temporary or episodic occupation. Region IIB was sterile and represented a phase of desertion. The archaeological materials found in the various levels confirmed these conclusions.

A more extensive vertical study was carried out by D. A. Davidson on the tell (a mound composed of many occupation layers) at Sitagroi, in northeastern Greece. This tell has a diameter of about 180 meters and a depth of about 11 meters, and it was occupied over the period 5400–2200 B.C. At least five major occupation phases had been recognized by the archaeologist Colin Renfrew.

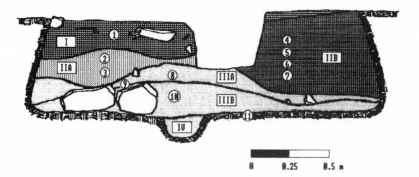

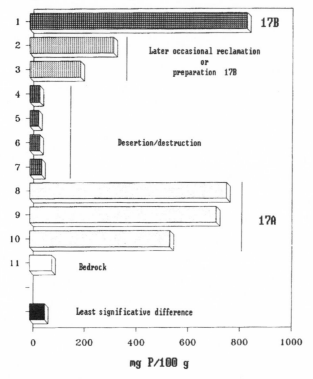

Figure 2.3. Top, *a cross section of Structural Complex 17 in the Spanish site of Polideportivo, Martos, Jaén. Roman numerals indicate horizons. Arabic numerals indicate locations at which soil was analyzed.* Bottom, *levels of phosphorus for the samples.*

One of the puzzles was the considerable depth of material between apparent occupation phases, up to 1.7 meters. Samples were taken at thirty-seven vertical positions in the tell (Figure 2.4), emphasizing the more homogeneous soil layers between occupation levels. Phosphate concentrations in the tell were uniformly higher, by factors of 4 to 15, than in comparison samples collected from nearby external areas that clearly lacked occupation. This observation sug-

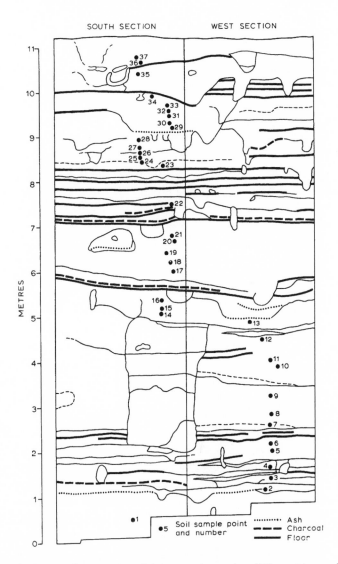

Figure 2.4. A vertical cross section of the excavation at the tell Sitagroi in northeastern Greece. Numbered points indicate where samples were taken for phosphate analysis.

gested that the tell had been continuously occupied. The archaeologists concluded that the tell grew as the result of occupation, and that the intermediate barren material most likely was debris from the collapse of houses. The houses presumably were constructed of local mud blocks and other alluvial material. On abandonment they decayed, melted away, and subsequently served as the base for the house of the next occupant. Phosphate levels tended to drop with depth, indicating a lower population density in earlier periods.

Prospecting with Other Elements

The chemist naturally strives to expand and improve on any method. Just as the use of more elements improved provenance determination of stone artifacts (Chapter 1), elements other than phosphorus also assist in archaeological prospection. The most likely candidates have already been mentioned: carbon, nitrogen, sulfur, sodium, potassium, and calcium, which like phosphorus are present in high levels in living organisms. Decomposition of organic materials is carried out very effectively by insects, bacteria, and fungi, with considerable loss of carbon, nitrogen, and sulfur. Sodium and potassium are highly soluble in water and hence may be removed fairly rapidly. More promising is calcium, which like phosphorus is a major constituent of bone and hence is produced by food residues and burials.

William I. Woods examined calcium as well as phosphorus to delineate the lost site of a French fort in Randolph County, Illinois. In the early 1700s the French constructed at least three forts, all named Fort de Chartres, in this county on the Mississippi River. The first two were of wooden palisades and after abandonment were lost. The third Fort de Chartres was built of stone during the 1750s and has left visible remains. In 1928 the U.S. Army Corps of Engineers carried out an aerial photographic reconnaissance of the area. Reexamination of these photographs in 1980 revealed a dark rectangular area about one kilometer from the remains of the stone fort. The dimensions corresponded closely to those indicated by historical documents for the first two forts.

Woods carried out a chemical survey of the site with three measurements: calcium, phosphate, and acidity. He collected soil on a 2-meter grid and studied several north–south transects in greater detail. On the transects, the phosphate and calcium concentrations peaked just inside the boundaries of the rectangular area, and the acidity dropped. Interestingly, the elemental concentrations dropped back down to the baseline over most of the central area of the rectangle. Woods interpreted these observations as indicating that the remains of the fence or palisade had brought about the high elemental levels, while the interior parade ground saw little human chemical contribution. The soil of the area is relatively acidic, so that the materials of the palisades and human activity served to neutralize the soil. In this study, calcium and acidity were used to confirm the conclusions based on phosphate, so that the case for having found the site of the lost fort was more believable. Moreover, John Weymouth carried out a magnetic survey of the site with complementary results.

Under certain circumstances iron also provides a means for prospection. Through chemical processes brought on by human activities, iron can concentrate in and around trenches, pits, postholes, and the like. Although invisible to the eye, the influence of iron can be enhanced by reaction with potassium thiocyanate (KSCN, commonly called rhodanide), yielding a bright red color. In this way features are visualized and excavation may be planned accordingly. Vegetation must be cleared before the liquid is sprayed, but this step is normally the case for any excavation. Some people experience skin irritation on contact with rhodanide. The general practice for disposal is simply to allow the stain to wash away with rain, in which it is soluble, although it then becomes an environmental contaminant. The colored patterns are recorded photographically for more extensive analysis later.

Analysis for Specific Human Activities

A distinction may be made between a survey, intended to obtain a general outline of a site, and a detailed examination of the chemicals in the soil, to determine what types of activities were carried out in a specific area. J. S. Conway's study of a native Welsh farm site of the Roman period illustrates how this kind of detail may be extracted from the chemical analysis of soil. The site of Cefn Graeanog in Gwynedd, North Wales (Figure 2.5), was a walled area containing several circular structures and a rectangular structure (E) that provided entrance to the hut group. Conway carried out a detailed analysis of the chemical makeup in three of the structures. Building C, in the northeastern corner of the group, was clearly a habitation site, as indicated by the presence of a hearth. Interestingly, phosphate concentrations were at a minimum in the middle of the structure and increased concentrically to the walls. This observation, in conjunction with analysis of the physical remains, led Conway to suggest that there had originally been a floor constructed of dung-tempered clay, or something similar, which had been eroded away by use in the center. Patches of finer material had survived close to the walls. In contrast, building E appeared not to have been used for habitation. The eastern half had remains of structures that suggested grain storage, while the western half was devoid of structures except for a posthole. The phosphate analysis of the western half showed a complex distribution (Figure 2.6), with high values along the central north–south axis and low values at the west end and along the central east–west axis. Conway suggested that this distribution indicated a physical division along the central east–west axis, supported by the post, dividing the area into two animal stalls. The drain at the east end provided an exit for excrement and other refuse and gave rise to high phosphate levels. The highest levels were found where the animals actually lived, while the low levels at the west end probably indicated a feature, such as a manger, that physically prevented the animals from contributing phosphate.

The team led by Craddock and Hughes combined phosphate analysis with artifact analysis at the previously described pre-Roman Iron Age site of Cat's Water. They found a number of round houses drained by deep ditches. The team argued that huts for humans should have more artifacts, whose distribution is patterned or nonrandom, with concentrations in specific areas of the hut. Huts for animals, on the other hand, should be uniformly richer in phosphate because of constant additions of manure. Indeed, they found that the huts with lower levels of phosphate had a higher incidence and a patterned distribution of artifacts, and hence were occupied by humans. The huts with higher levels of phosphate had an unpatterned distribution of finds and, interestingly, were usually located toward the center of the settlement. These probably held the animals.

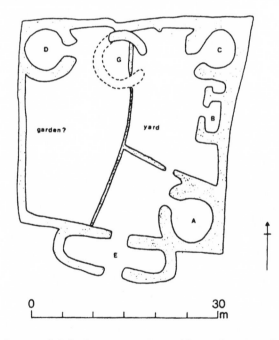

Figure 2.5. Layout of Cefn Graeanog in Gwynedd, North Wales. The arrow points to the north. The actual uses of structures C *and* E *were suggested by the patterns of phosphate deposition. Reprinted with permission of Academic Press Limited, London.*

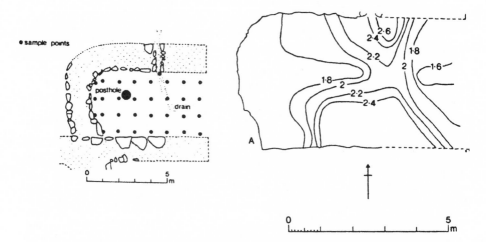

Figure 2.6. Left, *details of the western half of structure* E *(Figure 2.5) of Cefn Grae-anog.* Right, *levels of phosphorus in the western half of structure* E, *suggesting a division into two animal stalls along a central horizontal line. Reprinted with permission of Academic Press Limited, London.*

A site on Munsungun Lake in Maine, studied by Victor A. Konrad, Robson Bonnichsen, and Vickie Clay, illustrates the use of elements other than phosphorus for activity analysis. In particular, magnesium (Mg) is associated with both wood and certain minerals. These workers carried out a magnesium survey of the surface of the site on a 4-meter grid. They found high magnesium at a stone hearth and at probable ash dumps from a cabin. Locating hearths is useful not only for direct excavation but for possible sources of carbon for carbon-14 dating. The authors also used magnesium to locate areas associated with chipping and heat treatment of rocks, which can enhance magnesium levels. They found twelve areas of high magnesium not associated with high phosphate or calcium (which would have indicated normal habitation sites). One of these areas excavated indeed revealed a cluster of flakes, possibly an area for processing tools rather than habitation.

The floor of a building in the Roman site of Verulamium, not far from London, contained, in addition to the usual debris from a collapsed building (quartz sand, chips of flint or tile, pottery), numerous small black particles. Chemical analysis identified the material as iron oxide, specifically the form called magnetite, commonly produced as flakes from the action of a hammer on an anvil. The building thus could be identified as having been used by a blacksmith.

As early as 1965, S. F. Cook and R. F. Heizer constructed a model whereby concentrations of phosphorus and calcium could be used to calculate how many people lived at a particular site. William I. Woods used this model in his investigation of the Black Earth site in Saline County, Illinois. This is a very early (Middle Archaic) site, with carbon-14 dates from 5,645 to 4,860 years before the present. The approximately 785-year occupation produced an artificial mound or midden similar to a tell. From the total weight and volume of the midden, the duration of occupation, and the amounts of phosphorus in the midden and in control groups, Woods calculated that about 10 kilograms of phosphorus were added to the midden every year. Cook and Heizer suggested that each individual contributed about 1.2 kilograms per year, so Woods calculated that on average 9 individuals at a time occupied the site. A similar calculation based on calcium gave 70 individuals. The truth probably lies somewhere in between, and Woods suggested about 15. Such calculations have many approximations, including not only the rates suggested by Cook and Heizer, but also the constancy of the elemental concentrations over time without substantial loss through leaching.

Organic Residues in Soil

A human body contains many inorganic molecules, such as water, and uses many atoms other than carbon (essential minerals such as iron and zinc). Nonetheless, most of our nonaqueous body is composed of organic molecules, as is the case with all plants and animals. Prominent exceptions include the inorganic materials that make up bone, teeth, and shell. Decaying organic molecules are a defining constituent of soil and provide nutrients for plant growth. Human excrement and the residues from food preparation, storage, and waste contribute to the organic content of the soil. Consequently, organic analysis should reveal information about human presence and activities at archaeological sites. Organic analytical techniques have only recently been applied to these problems. Richard P. Evershed, Philip H. Bethell, and their co-workers have used fatty acids from food and steroids from excrement to map occupation and farming sites. These methods are considered further in Chapters 6 and 8.

Skeletal Silhouettes

Whether the objective is a general survey or an analysis for specific activities, soil analysis relies most heavily on materials that are the by-products of human activities—garbage, livestock, fertilizer, and excrement. Sometimes more tangible artifacts survive, and we will consider their analysis in the appropriate chapter in this book. Even when the object is entirely decomposed, it can leave a visual and chemical imprint in the soil as a contrast in color or texture, usually called a stain or silhouette. The silhouette of an entirely decomposed human body generates a stark impression for the viewer (Figure 2.7).

Does a skeleton for which there are no visible remains leave a chemical trace? In 1939 one of the most famous discoveries in British archaeology was made at Sutton Hoo, in Suffolk, of a Viking-era ship burial with extensive grave goods. Most puzzling, however, was the apparent absence of the principal character: There were no visible traces of a body in the burial chamber. Expecting to find enhanced phosphate concentrations as chemical traces of an entirely decomposed body, a team led by Harold Barker and Anthony Werner carried out analyses of the soil of the chamber. Although phosphate levels were generally high (an indication of the expected human activities), there were no specific areas of high phosphate concentration and certainly no pattern indicative of a body. A negative result is always inconclusive in science, so either there was no body or somehow it left no chemical trace.

More apparent silhouettes like that in Figure 2.7 generate more satisfying chemical results. High levels of iron and manganese explain the darkened nature of the stains. Unlike phosphate, which is a remnant of the original material, iron and manganese were present in the stains at levels well exceeding what was present in the original skeleton. Helen C. M. Keeley, in her study of silhouettes at the site of Mucking in Essex, suggested that manganese (and probably iron) is accumulated by microbial activity. The process is the

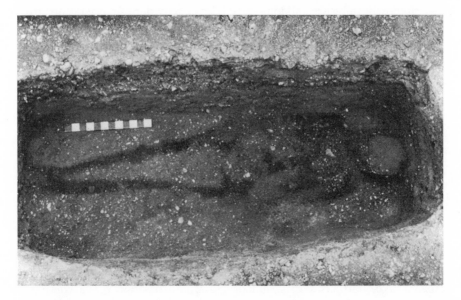

Figure 2.7. Iron-manganese stain as the remnant of a human body, in a burial in Mucking, Essex, England.

same as in the formation of desert varnish (Chapter 1). The organic matter of the body provides nutrients for bacteria, which pull in manganese from the surrounding soil. The author's laboratory carried out studies to test this hypothesis by analyzing soil around well-preserved skeletons from the Middle Woodland Period that were excavated from the Elizabeth site in west-central Illinois by Jane E. Buikstra. We found that iron and aluminum in the soil had indeed been depleted within a few centimeters of the skeleton, which in turn had gained significant amounts of these elements. In addition, calcium proved to be enriched in the soil and depleted in the bone. Thus the skeleton had lost calcium and gained iron and aluminum. The movement of elements between soil and bone or artifact is thus rather complex and depends on a host of factors, such as soil acidity, moisture levels, chemical makeup of the soil, and microbial activities. The chemical traces still help the archaeologist understand what happened to an object during burial and how the visible remnant in the soil can be interpreted.

Soil Stratigraphy

It is necessary to understand not only the structure of soil but also the order in which soil or sediment layers occur—that is, its *stratigraphy*. Geological processes that create layering include erosion, volcanic action, and sedimentation. They define the overall structure of many sites, on which human activities are superimposed. Intensive and long-term human activities can give rise to artificial mounds or tells, whose structure is composed largely of human debris. Sites with less intensive or more episodic occupations have layers determined geologically, with minor human perturbations. In either case, the layering is examined in detail by a wide range of specialists: not only chemists, but pedologists (soil scientists), geologists, climatologists, and biologists. Putting the whole picture together thus becomes a complex multidisciplinary operation. Frequently visual examination and qualitative physical analysis (texture, grain size, color) suffice to identify and differentiate layers and to compare them with similar layers at related sites. Chemistry can make an important contribution, particularly when the more qualitative methods prove insufficient. We have already seen how vertical sampling can indicate habitation from phosphorus levels. Chemical analysis also can provide information about how the soil or sediment was formed.

Sites dominated by volcanic activity or silting can be particularly subtle in their stratigraphic interpretation. The Yagi site, located on the southern end of Japan's large, northernmost island of Hokkaido, has two active volcanoes, Komagatake and Essan, within 30 kilometers of each other. During the last six thousand years, Komagatake has erupted at least ten times with significant ejecta. Some of these deposits, however, occur in only a small area. There are at least eighty-four other prehistoric sites within 10 kilometers of Yagi. Making comparisons of the various layers thus becomes very complex. Yagi had several occupations during Japan's earliest or Jōmon period, which may have seen the earliest manufacture of ceramics in the world. Ronald G. V. Hancock and co-workers sought to characterize the layers at Yagi by chemical analysis. They eventually examined almost two dozen elements, in particular the rare earth elements (REE). They succeeded in finding elemental patterns that enabled them to identify most of the Yagi volcanic layers, which in turn could be compared with layers at nearby sites to establish identity and contemporaneity.

A group led by Ralph O. Allen also used REE proportions to study Nile sediment layers at the Egyptian site of Hierakonpolis ("City of the Hawk" in Greek). This site may have been where leaders unified the Lower and Upper Egyptian cultures six thousand years ago. Allen and his archaeological co-workers used chemical analysis to characterize different Nile deposits according to their chronological order. In this fashion they were able to match deposits from two different locations in the area, El-Kap and Nekhen. Because one area could be dated to 3200 B.C. from pottery deposits, they could apply this date to the same layer found in other locations.

Soils and sediments in the form of clay provide the raw material for pottery and other ceramics. Chemical analysis can provide the key connection between the source of a raw material and the finished ceramic, just as it can for artifacts made from stone, but this is part of the story in Chapter 3.

Chemical prospection cannot compete with aerial photography in spotting features over a large area, nor with field walking in locating concentrations of artifacts. It is not likely that chemical analysis ever will be a dominant method for prospection. Its current strength lies in helping to sort out the uses to which settlement sites were put. It is possible that future progress will enable chemical analysis of soil to provide fingerprints for a wide range of human activities. Whereas phosphorus and calcium indicate the general presence of humans, other elements, in conjunction with artifact analysis, may delineate specific activities. Since wood contains high levels of potassium, a cooking area may contain high phosphorus and potassium. Since clay contains high levels of aluminum and iron, an area that was used for the intensive preparation or use of pottery may be high in these elements. Specific organic compounds also may characterize particular uses. Chemical fingerprinting of human activities by soil analysis is an evolving field whose importance and utility may increase as we improve our understanding of the relationships between specific activities and the elements and organic components in soil.

3

POTTERY

What Is Pottery?

Fire has been used for over a million years by humans and their ancestors as a source of heat, light, and protection. Its use to process food may have begun 100,000 years ago and later helped initiate our move from hunting and gathering to the more settled existence of farming. Fire has been used in the processing of stone for more than 50,000 years, particularly for hardening flint. At least 20,000 years ago figurines were molded out of earthy materials and hardened, in essence by being cooked. About 10,000 years ago humans began to form utilitarian devices out of materials from the earth and harden them with heat.

The creative use of fire in processing materials is termed *pyrotechnology*. Chapter 1 describes how heating gypsum or calcium carbonate provides the raw materials for plasters. With much higher temperatures, which required technical advances over simple wood fires, humans learned thousands of years later how to make glass and to smelt metal, as described in Chapters 5 and 7.

At about the same time that plaster was discovered, humans began to experiment with the properties of clay. This earthy material is widely available and probably first was appreciated for its ability to be worked when wet. Such a material is said to be *plastic*. After wet clay is shaped, it hardens on drying and generally holds its form. The discovery that heating such objects improves their utility by reducing their fragility may have been discovered when clay vessels were used as food containers during cooking.

Wet clay contains two types of water molecules. Those that are not bound chemically to the clay are lost through evaporation or heating to the boiling point of water (100°C). At much higher temperatures (about 600°C), clay begins to lose water that is bound chemically to the clay molecules. The silicon (Si), aluminum (Al), and oxygen (O) atoms (as aluminosilicates) are forced to adjust their structure to the loss of these water molecules, and the material loses its plasticity and hardens into soft forms of pottery. Above about 850°C, profound changes begin to occur in the aluminosilicate structure, a process that has been termed *vitrification* (from the Latin for "to make into glass").

Most primitive pottery is fired at about 900°C and falls into the category of *terra-cotta,* which is very porous and easily scratched. Heating at these temperatures also affects metal atoms present in the original clay, causing changes in the color. Iron (Fe) has a strong coloring effect in the range 900–1100°C, and the material produced, called red *earthenware,* is stronger and less porous than terra-cotta. At 1100-1200°C, calcium (Ca), in the form of lime, can bring out a cream color in earthenware. When earthenware is covered with a glaze, particular classes of materials such as *majolica* or *faience* are formed. Vitrification intensifies at higher temperatures, producing *stoneware* at 1200–1300°C, which is very strong and negligibly porous. Above 1300°C, highly vitrified and even translucent materials such as *porcelain* are obtained. Materials produced by heating clay or other nonmetallic minerals in general are called *ceramics.* Pottery includes any clay-based ceramic, whether a vessel, statue, tile, or brick.

The nature of the end product is determined not only by the firing temperature but also by the raw material. Three constituents go into the production of pottery. The first is clay. The normal weathering of silicate rocks leads to the formation of clay, which is made up not only of aluminum (Al), silicon (Si), and oxygen (O), but also of alkali and alkaline earth elements such as sodium (Na), potassium (K), magnesium (Mg), and calcium, as well as transition metals such as Fe. A variety of minerals are classified as clay, including kaolinite, illite, and montmorillonite, which are the most important for ceramics. Raw clay generally is crushed or ground and then wetted and sifted to remove coarse materials such as pebbles or twigs.

The second constituent is water. Dry clay is not workable, while clay containing too much water does not retain its shape. Within a narrow range of water content, roughly 25 percent, clay possesses plastic properties: it may be worked to a form that retains its shape. After that shape is attained, the water of plasticity is removed. The drying process results in shrinkage that, if not properly controlled, can lead to cracking or distortion of the product.

The third constituent is *temper* or filler. The addition to clay of materials without plastic properties allows water to evaporate more smoothly, minimizing shrinkage and preventing cracking. In some cases these materials may assist in vitrification. Potters have used a wide variety of tempering materials, including sand (quartz), limestone, shells, basalt, volcanic ash, mica, straw, dung, feathers, and even pottery shards (always crushed). Temper normally is added at the same time that the clay is brought to the right consistency by kneading with water. The material, called the *paste,* is then shaped, the surface is smoothed, and the object is allowed to dry to form the pottery body or fabric. Sometimes at this stage the object is dipped into a dilute suspension of clay in water to provide a thin layer, or *slip,* which provides a smoother and less porous surface. Moreover, the slip may contain a pigment to change the color of the surface, or it can serve as a ground for decorations.

The most primitive pottery probably was heated in an open fire, but enclosed structures called *kilns* provide higher temperatures and better control over the process. The two critical conditions are the temperature and the amount of oxygen. A simple wind-aided fire might attain a temperature of over 900°C, which is sufficient to produce terra-cotta or earthenware. More highly vitrified pottery, however, requires aids such as bellows or better constructed kilns. Under conditions with abundant oxygen from the atmosphere *(oxidizing conditions),* organic materials burn off as carbon dioxide, raising the vessel's porosity. In a poorly vented kiln *(reducing conditions),* oxygen is replaced by carbon monoxide or hydrocarbons, and the organic materials in the clay are left incompletely burned, imparting a dark color. Iron under these conditions also may lose oxygen to some degree and be reduced to black ferrous oxide (FeO).

The color of pottery thus is a function of several factors, including the original chemical constitution of the clay, the temperature of firing, the oxidizing properties of the kiln, and the surface treatment, such as the use of a slip. Blacks and grays are caused by a reducing atmosphere, either by charred organics or reduced iron. When conditions are oxidizing and iron is present, the color may vary from pink to red to orange, depending on the temperature of firing. Yellow and cream colors require temperatures above 1100°C and the presence of calcium. A white body, as in porcelain, also requires calcium and an even higher temperature.

Pottery shapes and uses probably had many preceramic models, including carved stones, reed baskets, gourds, and depressions in the earth used to contain fires (hearths). The earliest hand-building techniques for forming pottery included pinching (shaping a ball of clay), slab building (pressing slabs of clay together), coiling (using a long coil to build up the overall shape and pressing the coils together), and molding (pressing slabs into a mold). Wheel-building techniques exploit centrifugal force to increase the speed of work and to enhance the symmetry of the vessel. Whether formed by hand or thrown on a wheel, the vessel may be further trimmed, scraped, or beaten. By one such technique, an anvil or stone is placed inside the vessel while the outside is beaten with a paddle. In this way, for example, a rounded bottom may be produced.

A final step in working may be the application of a glaze over the entire surface of the vessel. Glaze is made of quartz sand and other materials and is formed by heating. The glaze may be fired at the same time as the clay, but more often it is applied to the fired product and finished in a second firing. In the two-step procedure the first firing is termed the *bisque,* which strengthens the body, allows most of the shrinkage to occur, and establishes a porosity that permits the glaze to adhere better to the body. Although glaze constitutionally is closely related to glass, it is included in this chapter as an inherent component of many types of pottery such as majolica and porcelain.

After decades of study of ancient pottery, archaeologists now can provide a wealth of information based on a visual examination of a piece's shape, style, color, decoration, and overall composition (presence of slip, glaze, and so on). Established changes over time can yield a reliable, although only relative, chronology. A characteristic shape or style may be associated with a particular source and therefore, if it is found at a distance from the source, indicate movement either of the pot through trade or the potter through migration. Shapes often reflect specific functions, so that the use of a pot may be established.

The physical scientist contributes information about what minerals are present and what elements constitute those minerals and other trace components. Such information can tell a lot about how and where the object was manufactured. Petrological examination or X-ray diffraction analysis can identify the minerals present, possibly from the original clay but more likely from the temper. In a classic study in the early 1940s Anna O. Shepard identified various rock tempers (andesite, sandstone, basalt) in Rio Grande Glaze-Paint pottery from New Mexico and associated them with local availability. In this way a pot could be identified with a specific location of manufacture. John Winter found that pre-Columbian Amerindians changed their temper when they moved from the Greater Antilles (Cuba, Hispaniola) to the Bahamas. Whereas quartz was used as a temper in the former locales, its absence in the Bahamas required the development of new tempers, such as crushed limestone or shell. Because trade occurred between the two locales, the specific site of manufacture can be determined from the composition of the temper.

Microscopy also provides information about the technology of manufacture. Mineral particles tend to be aligned in wheel-thrown pottery, but they are more nearly random in hand-built objects. Sometimes it is difficult to distinguish by eye the difference between a true slip and hand smoothing, but microscopy can locate the clear line of demarcation of a slip. Firing temperature and the type of atmosphere present may be explored by both microscopy and spectroscopy.

Chemical analysis has proved of considerable value in identifying the source of stone raw materials (Chapter 1). Since clay is weathered stone, the concept should apply to the raw materials for pottery as well. There are, however, numerous complications. For one, raw clay is processed through crushing and sifting, which remove certain components. For another, the potter may combine raw clay from two or more sources. The addition of temper also changes the levels of the original elements in clay and introduces new ones. Finally, the process of firing can remove volatile components. These problems are added to the usual ones: finding all possible sources of raw material, assuring that each source is homogeneous, distinguishing each of them, and hoping that during burial significant changes in composition did not occur. Because the chemical composition of a clay bed and a final piece of pottery can differ significantly, it is generally thought that it is not practical to identify sources of raw clay by elemental analysis. This depressing conclusion has been bypassed by chemists, who have discovered that elemental analysis indeed is very effective in identifying pottery groups. These groups may represent a specific kiln site, a single village, or a larger geographical entity that maintained sufficient commonality of raw clay, temper, and firing conditions to yield a product with a characteristic elemental fingerprint. The following sections illustrate how chemistry has been used in pottery analysis, from the earliest types of pottery to possibly the supreme example, porcelain.

The essentials of ceramics have been discovered or invented several times in world history—that wet clay has plastic properties, that adding other components improves the properties of the final product, and that fire hardens the object. Radiocarbon dates as early as $10,750 \pm 500$ B.C. have been measured for soil layers containing pottery in Fukui Cave, located in the northern part of Kyushu Island, Japan, corresponding to the Early Jōmon period. Although other techniques have yielded slightly more recent dates, these materials remain the oldest dated pottery. Probably an independent discovery occurred in southern Turkish sites (Anatolia) such as Çatal Hüyük and Beldibi, with dates in the range 8500–8000 B.C. In the New World the earliest dates are in the range 2500–2000 B.C., from Ecuador, Colombia, Mexico, and southeastern United States.

Early Pottery from the Eastern Mediterranean

Chemical analysis of pottery is most usefully applied to trading patterns when pottery of local manufacture is well characterized and then compared with other groups. The group of M. F. Kaplan, G. Harbottle, and E. V. Sayre applied these principles to a pottery type called Tell el Yahudiyeh ware that has been found in the eastern Mediterranean, including Egypt, Cyprus, Nubia, and the Levant, and that dates from the period 1750–1550 B.C. This is the Middle Bronze Age, which in Egypt corresponds to the Second Intermediate Period, when the Hyksos from the Levant displaced the Egyptian central government. Tell el Yahudiyeh pottery provides a chemical test of cultural ties between these groups in the areas where it is found. Figure 3.1 shows a typical example, although varieties are considerable.

The chemists carried out numerous analyses of clays and local wares from all the relevant areas. They found, for example, that local pottery from Ras Sharma in Syria had high levels of barium and chromium; that Nile sediment, or alluvium, and the derived pottery had high levels of manganese and scandium; that Sudanese vessels had high rubidium and cobalt; and so on. In this way they concluded that they could designate with confidence where a particular pot had been manufactured. They found that certain styles always were made in the Nile Valley, no matter where they were found. Similarly, other styles always were made in Levantine sites such as Syria. They concluded that there are two families of Tell el Yahudiyeh ware, one from Egypt and one from the Levant. Chemical analysis indicated that the portrayal of the Hyksos period by the later Egyptian texts was misleading, a typical case of the victors maligning the previous administration. Rather than being a dark age, this period enjoyed active trade among Egypt, the Levant, Cyprus, and Nubia. Not only did the objects move, but ideas and technology had to flow so that both the Levant and Egypt could support the manufacture of a single type of pottery.

One of the most extensive studies of this sort was carried out by Hector W. Catling and his co-workers on thousands of pieces of pottery produced by the

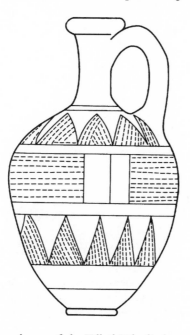

Figure 3.1. A typical early pot of the Tell el Yahudiyeh type found in the eastern Mediterranean during the period 1750–1550 B.C.

Minoan and Mycenaean civilizations. Originally thought to be associated with the heroic era enshrined by Homer in his epics the *Iliad* and the *Odyssey*, these cultures in fact flourished hundreds of years earlier. The Minoans from Crete (Figure 3.2) can make a claim for the first high civilization of Europe, lasting from about 3000 to 1200 B.C. (the final date is controversial, as we shall see). Their unglazed earthenware pottery surpassed that of Egypt aesthetically and eventually developed swirling representations of marine subjects such as lilies and octopi (Figure 3.3). The Mycenaean civilization was centered around Mycenae (Figure 3.2) and flourished from about 1600 to 1200 B.C. Their pottery was based on the Minoan model. At the height of the Mycenaean civilization, corresponding to the Late Bronze Age, Mycenaean pottery was the standard for the Greek world and beyond to Southwest Asia, Cyprus, Sicily, Italy, and Egypt. Such a widespread style implies extensive trade that may be examined by chemical analysis of the pottery.

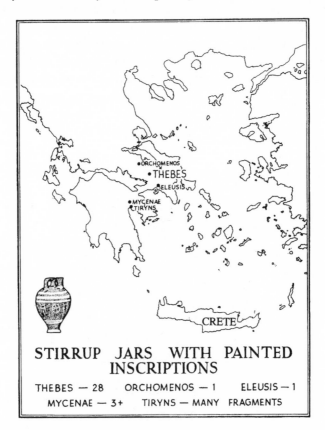

STIRRUP JARS WITH PAINTED INSCRIPTIONS

THEBES — 28	ORCHOMENOS — 1	ELEUSIS — 1
MYCENAE — 3+	TIRYNS — MANY	FRAGMENTS

Figure 3.2. The Aegean world of 3000–1200 B.C., including the find locations of stirrup jars with painted Linear B inscriptions.

The study by Catling of Minoan and Mycenaean pottery from the period 1400–1200 B.C. uncovered at least seventeen distinct groups, based on chemical analysis. By far the largest were Mycenaean pottery from the Peloponnese in Greece and Minoan pottery from Knossos on Crete. These were considered to be the two major production centers. One example of trade connections was provided by sixty Mycenaean shards found on Cyprus, forty of which corresponded to the group manufactured in the Peloponnese. Similar Peloponnesian-manufactured wares were found on the Aegean islands of Melos, Chios, and Rhodes, in Syria, and at the Tell el Amarna site in Egypt. These extensive exports from Mycenaean Greece contrast with much smaller exports from Minoan Crete, which must have decreased as a mercantile power at this time in the eastern Mediterranean.

Catling's group analyzed twenty-five so-called stirrup jars found at a post-1400 B.C. Mycenaean site at Thebes in Boeotia, Greece. These jars carried inscriptions in the early Greek script Linear B, which had been deciphered by Michael Ventris (Figure 3.2, in which the illustrated jar carries such inscriptions between the white bands). These jars proved to be key elements in the controversy over when Knossos on Crete was destroyed. Whereas its excavator Sir Arthur Evans favored a date of circa 1400 B.C., the historian Leonard Palmer and others favored 1200 B.C., which required a drastic alteration of the

Figure 3.3. A Late Minoan (ca. 1450–1400 B.C.) jar from Knossos, Crete, with a painted octopus design, height 18 inches, from the Ashmolean Museum. Reprinted with permission of John Johnson (author's agent) Ltd. © 1972 by Emmanuel Cooper.

Aegean chronology. As support for his argument, Palmer noted that some of the Theban jars carried Linear B names for known sites in Crete. The presence of these names on jars dating to after 1400 B.C. means that sites on Crete were still active. Chemical analysis of the stirrup jars by Catling and his co-worker Anne Millet eliminated a mainland source and originally suggested an eastern Cretan source, later revised to western Crete. Although consistent with Palmer's position, these chemical results were not definitive, as the dates for the production of the jars are uncertain. The controversy aside, Catling's study did show that Linear B was being used in Crete during the Mycenaean period of 1400–1200 B.C. and that there was active trading with the mainland.

We carried out chemical studies of Mycenaean pottery in conjunction with an archaeologist, Al Leonard, Jr. The focus of our study was Mycenaean pottery found in the late Bronze Age levels at the site of Megiddo (biblical Armageddon) in modern Israel, not far from Haifa. The objective of the study was similar to Catling's with the Cypriot shards. We found that none of our seventy-one pieces resembled the Peloponnesian group, which has high calcium. We found three large groupings, all with low levels of calcium. Although there was no exact match with the Catling groups, there were strong similarities with his group from Crete. We concluded that the Megiddo pottery came from at least three sources, with Crete as a strong possibility.

Our analyses used eight elements, and their relative levels were assessed and compared statistically by computer, as is now standard. Nonetheless, the scientist would like to be able to visualize elemental levels in plots. A simple xy plot can illustrate only two elements at a time. Three-dimensional plots are possible but still can handle only three elements. Statisticians have developed a means to visualize more than a dozen variables (elemental levels in our case) using the human face as the template. With this method, we set the shape of the upper face to reflect the level of aluminum; the shape of the lower face, calcium; the length of the nose line, iron; the curvature of the mouth, calcium (from a different spectroscopic quantity); the separation between the eyes, sodium; the size of the eyes, potassium; the position of the pupils, magnesium; and the slant of the eyebrows, oxygen. The method had numerous other unused dimensions, such as the vertical position of the mouth. As shown in Figure 3.4, these facial plots (the program is called FACES) not only give personality to the analysis, but provide an effective means to visualize all eight elemental variables simultaneously. Cluster 1 typically has a more rounded upper than lower face, relatively level eyebrows, and forward-looking eyes. Cluster 2 has a rounded head, left-looking eyes, and a frown. Cluster 3 has a more angular jaw and a rounded upper face. Smaller clusters, such as 7, show highly idiosyncratic expressions. High sensitivity to the details of the human face make it a useful multidimensional surface for plotting.

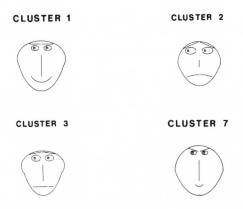

CLUSTER 1   CLUSTER 2

CLUSTER 3   CLUSTER 7

Figure 3.4. FACES plots of the chemical composition of Mycenaean pottery found at Megiddo, Israel.

Pottery from Greece

When one reads about a Greek vase in the popular press, it almost certainly is not of Mycenaean origin; nor probably is it of the Geometric style, which began in Attica around Athens about 900 B.C.; or even the black-figured style, which began in Corinth about 700 B.C. and continued for two hundred years (Figure 3.5). About 530 B.C. a new technique, called red-figured or red on black, was invented in Athens. It is quite likely that a single individual was responsible for the invention. This person sometimes is called the Andokides painter after the potter's colleague who decorated and often signed the product. The result was quite striking (Plate 3). The new technique permitted a more rounded human figure that is seen on many fifth-century masterpieces. The black figures had been achieved simply by painting them onto the previously smoothed surface, followed by a single firing under oxidizing conditions. The black material actually was neither paint nor glaze, but a slip of the same clay as the vase, only finer. What we might call industrial levels of pottery production were achieved in Athens at the time red-figured pottery became widespread. The potter's quarter or Kerameikos extended from the Agora to the Academy, about 1.5 kilometers. T. B. L. Webster has estimated that the total production of painted pottery in Athens may have numbered several hundred thousand.

The process for achieving red on black was much more complex than for the black figures, and defied chemical analysis until the 1940s. A ceramic chemist named Theodor Schumann had solved the problem of how the Romans had created the brilliant red color of Samian or Terra Sigillata ware associated with Arretium (modern Arezzo). He reasoned that Attic red-figured pottery might have been produced by a similar process. His assumption was that the process had to be simple to have been within the technical capabilities of sixth-century B.C. Greece. Microscopy showed that the glossy red or black surfaces had the same chemical composition as the body of the pottery. Thus no pigments, for example manganese, had been added to provide the black color. In fact, both red and black can be produced by different oxides of iron. A red color results from the presence of hematite (Fe_2O_3) produced under oxidizing conditions, whereas black results when lack of oxygen gives rise to a reducing atmosphere, stripping oxygen from iron, as in FeO or in $FeO \cdot Fe_2O_3$ (Fe_3O_4). The following technical procedure worked out by Schumann is often found in pottery texts, usually without giving him credit, as the invention that the Andokides painter made in 530 B.C.

According to Schumann, when the pot is formed and dried to the leather-hard stage, the figures to be red are outlined and the areas to be black are painted over with a slip. The slip was prepared by suspending the clay in water

Figure 3.5. Black-figured Athenian jar (ca. 560 B.C.), from the Metropolitan Museum of Art, showing women weaving. Reproduced with permission of publisher. © 1960 by Holt, Rinehart & Winston, Inc., and renewed 1988 by Glenn C. Nelson.

and selecting the finer, more slowly settling particles (a process called elutriation). To the suspension was added a material such as tannin to keep the particles dispersed and a basic or alkaline material such as wood ash. The alkali served as a *flux,* which is a material used to reduce the melting point. The slip then vitrifies at a temperature lower than for the body. The pot, with the slip in the areas intended to be black, is fired in three stages. (1) An oxidizing atmosphere and a temperature of about 850°C produces an entirely red pot. (2) A reducing atmosphere is produced by closing any air vents and adding green or wet wood to the furnace, thereby producing carbon monoxide, which removes oxygen from Fe_2O_3 on the entire surface of the pot, to produce carbon dioxide and the black forms of iron. At the same time, the lower-melting slip, intended to end up black, begins to vitrify, flows into surface pores, and seals that part of the surface. (3) At a slightly lower temperature, the atmosphere is made oxidizing again. On the unslipped surface, oxygen returns to iron, and the red color of hematite (Fe_2O_3) is restored. The vitrified slip, however, prevents oxygen from penetrating to iron in the body, so the slipped portion of the surface remains black. The key microscopic observation by Schumann was that the red surfaces always were relatively rough and porous, whereas the black surfaces were more fine grained, smooth, and impenetrable. He found that the glossy surface of the later Roman Samian ware similarly was no different chemically from the body. In this case a slip was applied to the entire surface of the vessel, which was fired in an oxidizing atmosphere at about 950°C, at which the slip vitrified to a lustrous, glossy surface. The higher temperature was required because the fluxes used by the Romans were not so effective as those for the red-figured ware.

Knowledge of this technique means that the presence of pigments to achieve the black color can be taken as evidence of forgery. A statuary set of enormous Etruscan warriors in the Attic style at the Metropolitan Museum of Art were shown to be modern forgeries in this way.

Glaze

All the materials considered to this point have been unglazed terra-cotta or earthenware. A glaze may be considered to be any glassy material coating the surface of a ceramic body, but there are differences between glaze and glass. In particular, glaze (1) contains a larger proportion of aluminum as alumina (Al_2O_3), which permits glazes to be fired at higher temperatures than glass (important for stoneware and porcelain); (2) possesses an affinity for the pottery body, which of course also contains aluminum; and (3) contains aluminosilicates, whereas glass (as discussed in Chapter 5) is a purer silicate. These three attributes are interrelated.

The discovery of glaze may have occurred by accident or by a series of developments. Glaze never was discovered in the New World (nor was the potter's wheel), but it did occur there by accident. Jane E. Buikstra, with her co-worker Ann Magennis, while excavating a crematory called the Helton site in the Lower Illinois River Valley, uncovered a skull and torso with a gray-green coating that bore every resemblance to glaze. The author's group analyzed the substance and found that it contained all the usual chemical components of glaze—including silicon, aluminum, calcium, and potassium—except lead. Instead it contained large amounts of iron, which could act as a fluxing agent. The scenario we envisaged was a cremation that included sprinkling sand and red ochre in generous quantities on the body (red ocher is a form of hematite, Fe_2O_3, used as a pigment) and firing with hardwood, such as elm or hackberry. The coincidence of materials in a hot fire could produce a natural glaze.

Widespread use of glaze on pottery first occurred in the Old World in Mesopotamia after 2000 B.C. A fifth-millennium B.C. Egyptian steatite figurine has been found with a glazed surface, and glazed tiles were known in Egypt by the fourth millennium B.C. and in China by 3000 B.C. These types of glaze were not suitable for the curved surfaces of pots, because of the different shrinkage properties of the body and the glaze. Mesopotamian and Egyptian glaziers experimented with numerous minerals as additives to glass, both as fluxes and as colorants. They found that lead minerals provided the ideal fluxing properties, and, in addition, controlled shrinkage and gave a satisfying luster to the product. A clay tablet from northern Iraq dating to 1700 B.C. provides a recipe for a lead glaze, and glazed pots have been found in northern Syria from the period 1700–1400 B.C.

There have been numerous types of glazes, of which lead glazes are the largest early group. The toxic properties of lead were not appreciated until the nineteenth century. Problems exist in both the manufacture and the use of lead glazes—for example, acid-containing materials such as citric fruit juices can leach out lead. There are numerous other types of glaze. Alkaline fluxes (including soda ash and borax) may be safer than lead, but have less desirable properties of expansion and contraction during firing and cooling. Stoneware and porcelain have high-temperature glazes made of flint or the white clay called kaolin, and fluxes of limestone or feldspar (an aluminosilicate mineral). Bristol glaze is another high-temperature material, which contains a large amount of zinc oxide. Salt glazing was developed in Germany around A.D. 1400 for stoneware. Common salt (sodium chloride or NaCl) is added to the kiln toward the end of the firing. The chloride part volatilizes as gaseous chlorine, but the sodium part is incorporated into the clay surface as a glassy silicate. The result, depending on the clay body, can be either smooth or orange-peel rough. Today salt glazes are found in stoneware crocks and sewer pipes, but saltware was very popular in central Europe from about A.D. 1500, as containers for ale. Whereas malt liquor was not a popular drink, the addition of hops starting around A.D. 1500 produced ale and later beer (Chapter 6). These drinks became exceedingly popular and created an enormous market for salt-glazed mugs and other containers for the new drink.

From Majolica to Jasper

The reasons for using a glaze are both utilitarian and aesthetic. On the one hand the smoother surface improves the object as a container for liquids and as a cooking vessel; on the other hand the glaze provides an excellent background for decorations. Lead-glazed pottery probably was in continuous production in Italy from the Roman period. During the Renaissance, however, tin-glazed pottery came into wide use. Although glazes containing tin oxide as well as lead as a flux have been found on Assyrian brick panels as early as 900 B.C., such glazes were not used by potters until the ninth century A.D. The presence of tin produces a deep white, opaque finish that is ideal as the background for painted ornamentation. Islamic potters of Samarra on the Tigris River during the period of the Abbasid caliphs (ninth century) invented or rediscovered true tin glazes. By 1302 the details of its preparation were described in a recipe book by Abu'l Qashim, associated with Kashan in central Iran.

Although the use of tin-glazed pottery in Europe during the Renaissance has been considered to be an independent discovery, it is likely to have been introduced from the Islamic world. Italian tin-glazed pottery has been called *majolica* or maiolica after the Spanish island of Mallorca, which served as a way station for the importation of the material from Spain. This Hispano-Moresque ware may have derived from the Islamic Moors, who had occupied parts of Spain. In the same fashion the French imported this glazed pottery from Faenza, Italy, and called it faience. Later, Delft in the Netherlands developed an active industry of the same material, although with decorative imitation of Chinese porcelain, and it became known as delftware. There also may have been influence in Italy directly from southwestern Asia during the period of the Crusades. The earliest firmly dated tin-glazed pottery in Italy is from Assisi, where majolica tiles were used in 1236–1239 to construct the high altar of the Upper Church of Saint Francis. Earlier dates have been found in Cyprus and in Corinth, Greece. Thus influences on Italian majolica came from both the west and the east.

Certainly the technique was widespread within Italy during the thirteenth century. It has been called archaic majolica with decoration in browns and greens in central and northern Italy, and proto-majolica with many colors in southern Italy and Sicily. Two-handled majolica jars (Plate 4) were very common, and, particularly from Faenza, the albarello ("little tree," based on the shape of bamboo) was popular as a container for medicine. By 1550 production was widespread throughout Italy and by 1600 had spread to northern Europe. Luca Della Robbia and later generations of his family used tin-glazed materials for the production of sculpture, since it gave the impression of marble (Figure 3.6). Decoration could be scratched or incised into a slipped object, which then was fired with a transparent yellowish lead glaze to give *sgraffito* ware, which is not majolica.

Figure 3.6. Terra-cotta sculpture with lead/tin glaze, by Luca Della Robbia, of the madonna and child (fifteenth century), from the Metropolitan Museum of Art. Reproduced with permission of the publisher. © 1960 by Holt, Rinehart & Winston, Inc., and renewed 1988 by Glenn C. Nelson.

Majolica moved to Spanish America soon after the discovery of the New World. Between 1504 and 1555, according to Spanish records, some 2,805 ships sailed from Seville and the Canary Islands to ports in the New World, carrying cargoes that included large amounts of majolica. All the raw materials, however, were available in Mexico, and at some point in the sixteenth century local production began. By the next century the Mexican city of Puebla had become a leading center for the production of majolica. The critical crossover from importation to local production has been examined by Jacqueline S. Olin, M. James Blackman, Garman Harbottle, and Edward Sayre. The construction of the Mexico City subway provided thousands of early shards, and repairs on the Metropolitan Cathedral provided samples that had been sealed under the floor in 1573. Other early samples came from Santo Domingo and Isabela in the Dominican Republica, Nueva Cádiz in Venezuela, and Fig Springs in Florida. These were compared with pottery from a contemporary Carthusian monastery in Jérez, Spain, and with indigenous Aztec pottery.

Elemental analysis demonstrated that majolica and other types of pottery produced in Mexico were easily distinguished from Spanish pottery, primarily on the basis of the levels of cerium, lanthanum, and thorium. In collaboration with Marino Maggetti, Olin also showed petrographically that pottery made in Spain contained temper from sedimentary sources, but pottery made in Mexico, not surprisingly, contained temper from volcanic sources. She went on to demonstrate that the Mexican sources could be further divided into production in Puebla and production in a second center, which may have been Mexico City. Figure 3.7 shows the excellent separation they observed based only on the two elements, chromium and iron.

Olin and co-workers also examined possible sources of raw material for the glaze on Mexican majolica, as it could have come from either Spain or Mexico. The lead in the glaze is present as several isotopic forms, whose proportions reflect the geographic source of the raw materials. They studied both glaze and ore samples and obtained an excellent separation based on lead-207 versus lead-208 between Mexican lead sources and two different Spanish sources (Figure 3.8). Thus majolica made in Mexico used a glaze whose lead had been extracted from local deposits. The production of majolica was entirely self-sufficient in Mexico by about the middle of the sixteenth century.

By the eighteenth century, Chinese porcelain and its imitations had become very popular in Europe. Tin-glazed earthenware like majolica was considered to be of lower quality. Because of the expense of porcelain, particularly during times of political unrest in China, European potters sought alternative earthenware formulations that resembled it. The Englishman Josiah Wedgwood is the primary person associated with these advances in glazed earthenware. Using trial and error experiments, he discovered in 1763 that the addition of kaolin

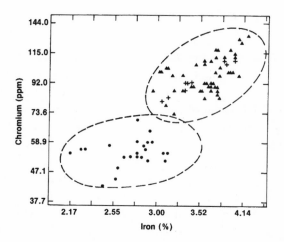

Figure 3.7. The proportions of iron plotted against chromium, showing the separation of samples produced in Puebla, Mexico (triangles, archaeological samples; pluses, modern samples) and those probably produced in Mexico City (filled circles). Reprinted with permission. © 1989 by the American Chemical Society.

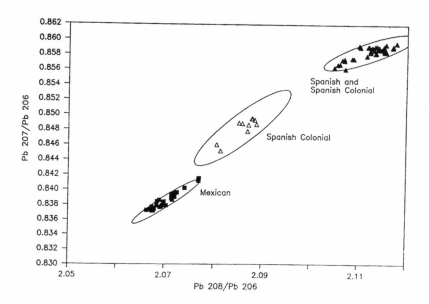

Figure 3.8. Use of lead isotopes to separate majolica glazes with Mexican sources (filled squares) from those with Spanish sources (triangles).

and feldspar produced a creamware (a cream-colored earthenware) with a stronger body than that of majolica. He used a clear lead glaze that completely replaced the salt and tin glazes of the period. This pottery has been called Queen's Ware, since it was adopted by Queen Charlotte for her tea service.

In 1774 Wedgwood discovered that the addition of barium sulfate could produce a pure white, porcelainlike stoneware, which he called *jasper*. The intense white, unglazed surface was easily stained with metallic oxides to yield a variety of colors, including yellow, lilac, and blue. When the colored background was contrasted with raised white decoration in low relief, the result was a classic form, which reached its apex in Wedgwood's copies of the ancient glass Portland vase (Figure 3.9).

Porcelain

Porcelain is generally considered to be the potter's highest achievement, both technically and aesthetically. Its discovery occurred in China some time before A.D. 700 and required advances in raw materials, firing techniques, and glazing. Recent excavations in the northern China provinces of Xian and Anyang uncovered white porcelains that dated to the late sixth century. Yanyi Guo, however, sets the invention of porcelain to an even earlier date, the Eastern Han Dynasty (A.D. 25–221), when porcelain stone was discovered and first used at the Yue kilns in Zhejiang. Early porcelain, however, is generally associated with the Tang Dynasty (A.D. 618–906). Materials that are white, translucent, strong, and sonorous on being struck were fully developed during the Song (Sung) Dynasty (A.D. 960–1279, Figure 3.10). During this period and the later Ming Dynasty (A.D. 1368–1644), the imperial kilns at Jingdezhen (Ching-te-chen) in Jaingxi province dominated production and trade. They were ideally situated for access both to raw materials and to river transportation. During the Ming Dynasty potters developed *underglaze* painting (in which decorations are applied before the glaze), which superseded monochrome glazes, as in Figure 3.10. From fourteenth-century Persia the Chinese obtained access to cobalt blue pigments that were superior to local cobalt ores contaminated with manganese. These blue-and-white wares became the epitome of porcelain for many Europeans (Plate 5).

Porcelain is composed of white clay (kaolin) and a feldspathic rock (petuntze), and is fired at or above 1300°C. At these temperatures the materials vitrify and the particles fuse to form the thin white walls of extraordinary strength and resonance that make up porcelain. When the Venetian Marco Polo returned from his travels in the Far East in 1300, his tales included descriptions of the white pottery that he called porcella (literally "little pig," a term derived from a highly valued shell used for currency in East Asia). For centuries Europeans failed to reproduce the product because they could not locate appropriate clay raw materials. As the body or fabric of pottery (under any slip or glaze) is called the paste, and as the earthenware then available in Europe was less strong than porcelain, the native European pottery of the time is called soft paste. Soft-paste imitations of porcelain had been achieved by the sixteenth century, and soft-paste factories—for example, at Sèvres in France— produced excellent products for several centuries.

Figure 3.9. White-on-black jasper reproduction of the Portland vase (the Hope copy, height 10 1/8 inches, 1793), from the Wedgwood Museum. The reclining figure is Thetis, accompanied by Hermes and Aphrodite. Courtesy of the trustees of the Wedgwood Museum, Barlaston, Staffordshire, England.

It was a German alchemist named Johann Friedrich Böttger who finally achieved the European preparation of porcelain. The traditional goal of an alchemist was to convert base metals such as lead into gold. Böttger made claims along this line that came to the attention of Friedrich I of Prussia. To escape from possible involuntary servitude for the purpose of creating gold, Böttger fled in 1701 to Wittenberg. Unfortunately Augustus, the Elector of Saxony, had similar designs on his gold-making pretensions, took him prisoner, and set him up with a laboratory in Dresden and later in Meissen. He of course had no success. In 1707 a nobleman named Ehrenfried Walter von Tschirnhausen asked permission from Augustus for Böttger to work on a project to develop porcelain. Baron von Tschirnhausen had been interested in

Figure 3.10. Porcelain vase with a pale-blue glaze (Sung Dynasty). Reproduced with permission of HarperCollins. MM-1096-65. © by John Bartholomew & Son, Edinburgh.

the economic advantages of the local production of porcelain for many years. Böttger inscribed over his door *Gott hat gemacht aus einem Goldmacher einen Töpfer* ("God made a potter out of a gold-maker") and set to work. The legend is that kaolin beds were discovered when Tschirnhausen wondered whether the mineral powder used by his valet to dust his wig could serve as the raw material for porcelain. The source was sought and found in Aue in southern Germany, and the material was kaolin. Böttger also included in his formula for porcelain a type of feldspar (similar to petuntze) called *Siebenlehnerstein*. In January 1708 he carried out a firing that produced a true porcelain product, and he took the occasion to also inform the king that he was not able to make gold. A royal porcelain factory was established in Meissen in 1710.

Böttger's achievement, in complete ignorance of Chinese methods and without training as a ceramicist, changed the face of the ceramic industry in Europe. The secret could not be kept, and porcelain factories soon sprang up all over the continent. Kaolin beds were discovered at Saint-Yreix, France, in 1766, leading to the creation of a factory at Limoges and the later conversion of the royal factory at Sèvres to hard-paste porcelain manufacture.

Soft-paste porcelain imitations had been produced in England from 1742 at Chelsea, from 1744 at Bow, and from 1751 at Worcester. The Chelsea porcelains resembled those from France, contained sand, clay, and flint, and have been called glassy porcelains. The Bow porcelains contained sand, clay, and bone ash, whereas those of Worcester contained sand, flint, and soapstone. William Cookworthy, a Plymouth chemist, discovered the ingredients for true, hard-paste porcelain in the same way as had Böttger, the kaolin component in England being called china clay and the feldspar component china stone. He patented the process in 1768 and continued the manufacture in Bristol. The components became more widely available in 1789 when the patent expired. About 1800 Josiah Spode combined the raw materials for true porcelain with bone ash to create a white, translucent, strong product that, though soft-paste, closely resembled porcelain. Spode's products became known as *bone china* and eventually dominated the English porcelain markets. They had to compete, however, with the less expensive earthenware creamwares of Wedgwood and his followers.

The description of how Chinese porcelain originally was made comes from knowledge of present-day and historical practices, but chemical and mineralogical analysis is required to determine whether changes have occurred in these practices over time. Current practice in China, and those rediscovered by the Europeans Böttger and Cookworthy, require the combination of kaolin clay and a feldspathic stone, called respectively china clay and china stone in England or porcelain clay and porcelain stone (petuntze) in China. Kaolin has a high aluminum content with little flux and by itself is difficult to vitrify below 1400°C. Alone it is more suitable for making soft-paste earthenware. Porcelain stone thus serves as a flux and assists in vitrification. The best porcelain contains about equal amounts of kaolin and porcelain stone, and lower grades have higher proportions of porcelain stone.

Michael S. Tite and his co-workers analyzed a range of early Chinese porcelains, both the body and the glaze, as well as candidates for the raw materials. They focused on *yingqing* porcelain, representing early types produced at Jingdezhen from at least the eleventh century, and blue underglaze porcelain representing later types. All the materials were actually from the Yuan Dynasty (1260–1368). Both they and Yanyi Guo found that porcelain stone was made up of quartz and a form of mica called muscovite or sericite. (Tite also found albite, a form of feldspar.) Kaolin contains these components in addition to the mineral kaolinite. Porcelain stone proved to be plastic like kaolin and fusible like feldspar, so it possessed all the properties needed to make porcelain. Guo in fact considers the discovery of porcelain stone in southern China to be the major factor in the creation of porcelain. The analysis of early porcelains such as many of the yingqing type indicate that only porcelain stone was used as the raw material—that is, kaolin is absent. Guo points out that the plasticity of these early materials was poor and depended on a very narrow firing temperature, so that the yield of misshapen vessels was high. During the Yuan Dynasty larger vessels were in vogue and porcelain stone alone did not serve. At this time kaolin began to be added to the porcelain stone to form a blend that was superior to the separate components. The blend increased the aluminum content to above 20 percent. Analyses indicated that during the Yuan Dynasty 10–20 percent kaolin was added, with the ratio (and the aluminum content) continuing to increase during the Ming and the Qing (1644–1912) dynasties.

Porcelain glaze is closely related to the material used for the porcelain body. Analysis by Tite and others suggested that the glaze raw material, called glaze stone, in fact was simply porcelain stone to which limestone had been added. (Fired limestone sometimes is called glaze ash in this context.) Guo indicates that glaze stone was manufactured by crushing and powdering porcelain stone to a finer texture.

The history of porcelain illustrates much of the ten-thousand-year evolutionary development of pottery and how chemists have come to understand it over the last fifty years. The choice of raw materials moved from crude clays and tempers to the sophisticated mix of kaolin and porcelain stone that produces the superb body of porcelain. Improvements in kiln technology enabled the high temperatures to be reached that are required to achieve extensive vitrification for stoneware and porcelain, as well as to select the cream and white colors obtainable only at higher temperatures. The ability to manipulate the oxygen content of the kiln atmosphere led to the complex coloring of Greek pottery. The creation of a thin-layered surface slip provided not only a base for decorations but also a material whose vitrifying properties could lead to red-figured Greek pottery and to the glossy Terra Sigillata of the Romans. The development of glazes added luster to pottery and enabled the creation of complex and highly colored decorations, as in majolica, creamware, and porcelain. Modern chemical analysis has led not only to an understanding of these developments, but has also linked them to the sources of raw materials and to pottery trade. This pattern of technical development and chemical tracing has parallels in the evolution and understanding of glass (Chapter 5) and metals (Chapter 7).

4

C O L O R

Dyes and Pigments

"I had almost forgotten these purple snails and their stench. The old city stank of their corruption as one came near, and the alleys made you hold your nose. Yet the purple of paving stones on every street with a dyer's shop was so bright, it hurt the eye. One could even see the sky reflected in that wet purple." In this way Norman Mailer imagines the city of Tyre in his novel *Ancient Evenings*. Set during the lifetime of the Egyptian pharaoh Ramses II in the thirteenth century B.C., the novel illustrates how people for millennia have engaged in the production of color-bearing materials, first by extracting them from natural sources such as snails and eventually by combining several raw materials to produce substances unknown in nature.

Our language has become rich with words to describe our perceptions of color—azure, mauve, cyan, fuchsia, khaki—which can evoke fond memories of sunsets and flower bouquets, or decidedly unfond memories of basic training. The receptors in our eyes are sensitive to a small window of the electromagnetic spectrum: light with wavelength in the range of about 400 to 800 nm. (A nanometer, abbreviated nm, is six orders of magnitude smaller than a millimeter.) When the full range of visible light returns to our eyes from an object, we perceive it as white; when no light returns (because the object absorbs it all), we perceive it as black. When an object absorbs some wavelengths of light and therefore returns only a portion of the visible range back to the eye, we perceive the object as colored. The absorbed and the perceived colors are said to be complementary. For example, when violet light is absorbed in the range 400–440 nm, we see a greenish-yellow color, and when red light is absorbed in the range 600–700 nm, we see a bluish-green color. Shades of gray are not true colors, as they are mixtures of black and white. When mixed with colored dyes, gray makes them flatter.

Light of a particular color may be obtained by mixing any number of pure colors, or *hues*. Figure 4.1 demonstrates how pink-appearing light can be produced by three different mixtures of hue. In A, pure orange light is diluted with white light to give pink. Alternatively, in B, pink is produced by mixing red and cyan (blue-green) light. As a third of any number of possibilities, in C, pink comes from the combination of red, green, and violet light. A scientist trying to characterize the color of a glaze or textile thus prefers to examine the light absorbed rather than reflected by an object. Reflected light would be pink for all the examples in Figure 4.1, but the absorbed light, as viewed in an *absorption spectrum,* would respectively have one, two, or three components. The observed and absorbed colors are said to be complementary. Thus if an object absorbs blue light, it is perceived to be the color of its complement, yellow. An example of an absorption spectrum may be found later in this chapter in Figure 4.10.

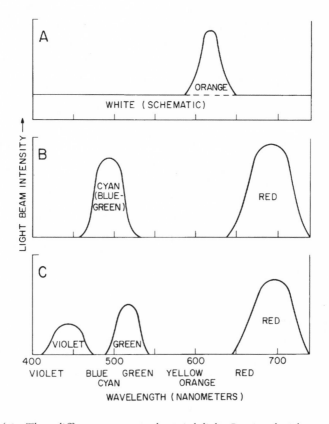

Figure 4.1. Three different ways to produce pink light. Reprinted with permission of John Wiley & Sons., Inc. © 1983 by John Wiley & Sons, Inc.

Light enters the electron cloud of the atoms and molecules that make up an object and is absorbed at specific wavelengths. The nonabsorbed light can be reflected back or transmitted on to the viewer. The object appears colored because this light no longer contains all the spectral wavelengths. An opaque object reflects light but allows none to pass through it. A transparent object lets all the light pass through. Since no light is absorbed or reflected, the object appears clear and colorless. A translucent object absorbs or reflects light but lets some pass through. As the object absorbs light, the transmitted light that comes out on the other side is colored.

Coloring materials are usually classified as *dyes* or *pigments*. Dyes are chemically bound to a substrate such as a textile, whereas pigments require another substance, called a *binder*, to help them adhere to the substrate. The human body may have served as the first substrate for dyes, but textiles or fabrics have been the most common over the ages. In addition to textile fibers, dyes have been applied to paper, wood, food, cosmetics, fur, and leather. The process whereby a dye is applied to a substrate depends largely on the chemical properties of the dye. The term *direct dye* refers to materials that are able to bind chemically to the substrate itself. The binding forces often are relatively weak, so that the dye may be slowly lost. Dyes that are easily lost are said to be *fugitive*.

Dyes that are poorly soluble in water are not easily applied to fabrics. *Vat dyes* are generally water-insoluble materials that may be chemically altered to a soluble and often colorless, or *leuco*, form (or *leuko*, from the Greek word for "white"). For example, some insoluble dyes contain a carbonyl group (carbon attached to an oxygen atom by two bonds, represented by C=O). Chemical conversion of C=O to the singly bonded form, the hydroxy group (C–OH), generates a leuco substance that dissolves in a strong aqueous base such as lye (sodium hydroxide, $NaOH$). The process was traditionally carried out in a vat or tub. Once the leuco form was bound to the fabric, treatment with air regenerated the original dye form with the carbonyl group. Such a process was designed to yield a very stable or *fast* color.

Often a dye may be attached to a chemical intermediary that itself is able to bind to the substrate. The intermediary is called a *mordant*, from the Latin word *mordere*, meaning "to bite." Almost all mordants contain metal atoms such as aluminum (as alum), chromium (as chrome or potassium dichromate), natron (sodium carbonate), iron, or tin. The mordant may be placed on the fabric before dyeing, it may be present in the dye bath, or it may even be applied after the dye. The chemical bonding usually is very stable, so the fastness properties of mordant dyes are excellent. The presence of the mordant may alter the color of the free dye. For example, the dye cochineal appears red with an aluminum mordant, brown with an iron mordant, and gray to beige with a copper mordant (Plate 6). Dyeing of materials other than textiles often requires special procedures. Dyes for paper are added to the pulp, and dyes are applied to hides or skins after tanning by heating and tumbling them in drums. The earliest known examples of mordant dyeing were with wool, in eighteenth-century B.C. Mesopotamia.

Almost all dyes are organic materials—that is, they belong to the large class of chemicals whose structure is based on the element carbon. On the other hand, most early pigments did not contain carbon and hence were inorganic. Water-soluble organic dyes, however, can be converted for use as pigments by embedding them in a solid substrate such as powdered chalk or kaolin clay. The resulting material, called a *lake,* would be powdered and used in the same way as inorganic pigments. Today, however, there are entirely organic as well as inorganic pigments.

Because pigments do not dissolve in most media, such as water, they require a binder in order to be applied to a substrate. No chemical reaction actually takes place between the pigment and the binder. The binder serves only to incorporate the pigment onto the surface of the substrate, providing another contrast with dyes, which permeate the substrate fully. Normally, binder and pigment are mixed mechanically and then applied together. For underglaze decorations on pottery, however, the pigment is applied first and the transparent glaze, serving as the binder, afterward. In *fresco* painting, the binder is formally applied first in the form of damp plaster prepared from lime (calcium oxide, CaO). The pigments, mixed with lime and water, are then applied to the damp surface. Lime reacts with the atmosphere to form calcium carbonate ($CaCO_3$, the same material as limestone), so that the pigments are essentially encased in stone.

In general, binder and pigment form two separate phases—that is, like oil and water they are unable to mix (they are *immiscible*). On the other hand, a dye forms a single phase that can transmit, absorb, or reflect light. Pigments also can transmit, absorb, or reflect light; additionally, the boundary between pigment and binder can scatter light. Because scattered light can have a different wavelength from reflected light, the perceived color is influenced both by reflection and by scattering. The scattering of light depends on the size of the pigment particles and on the difference in the abilities of the pigment and the binder to bend, or refract, light. The extent that light bends on going from vacuum to a substance such as the pigment or binder is the *index of refraction* of the substance. If the particles of pigment are too fine or too coarse, the perceived color may be altered adversely. Loss of color is termed *desaturation,* which may occur when a brightly colored material is powdered.

It is interesting to note that there are many white pigments but no white dyes. By definition dyes absorb some light and hence convey color. When a single phase like pure water in a glass absorbs no light, it is clear and colorless rather than white. Only by scattering light can a substance appear white. Thus a nonscattering, nonabsorbing substance, such as a varnish that covers a substrate, appears clear and colorless and does not alter the color of the substrate. When all wavelengths of light are scattered, the object appears white. In this way a white surface pigment can hide colors underneath it.

Some pigments such as ultramarine are intense absorbers of light even though they have a low index of refraction. Pigments with high indexes of refraction—in contrast to the binder, with a typically low index of refraction—scatter light strongly and have good covering power. Vermilion, red lead, and chrome green, for example, have high refractive indexes. White pigments with good scattering power are added to poorly absorbing, low refractive pigment in order to enhance its ability to hide other colors.

To serve as a *paint*, pigments are mixed with a carrying agent or *solvent* as well as a binder. The purpose of the solvent is to keep the mixture fluid until application is complete, after which it evaporates. Chalks and crayons contain pigment and binder but no solvent. The binder in *oil paints* is linseed oil (from flax), tung oil (from the tung tree), or any of a variety of other (usually) plant products; the solvent is an organic material like turpentine. During the drying process the solvent evaporates, and double bonds between carbons (C=C) in the binder combine with other binder molecules, or polymerize, to form larger and larger molecules (polymers). This drying or hardening process is accelerated by light and air. Oils that lack double bonds cannot polymerize and are called semidrying (cottonseed oil, sesame oil) or nondrying (olive oil, coconut oil). They are not suitable as binders. To prevent drying oils from cracking or decaying, the finished product often is covered with a clear film called a *varnish* that provides a seal. Most modern paints use synthetic materials called *acrylics* as binders, but the principle is the same, as the acrylic material (such as vinyl acetate) contains double bonds that polymerize upon application. Synthetic formulations are designed to dry quickly, to use water as the solvent (a *latex paint*), to be flexible, and to have superior hiding power. Their stability over centuries, however, is not known.

Tempera paints use an oil-water *emulsion* as the binder. An emulsion is a stable mixture of two normally immiscible liquids. The process of mixing oil and water requires an emulsifying agent such as gum arabic (from the acacia tree), glue, or egg yolk, or else the two materials will separate. In *watercolor* paint, the pigment is ground very finely and suspended in water with a binder such as gum arabic. *Inks* are liquids used for writing or drawing. The coloring material may be either a pigment or a dye, together with a binder, such as a gum or a glue, and a carrying agent (or *vehicle*), such as water.

The Earliest Coloring Materials

The first cultural use of color is lost in the depths of unrecorded history. It may have taken place as early as half a million years ago or more and involved body painting. Within the last hundred thousand years both Neandertal (Neanderthal) and Cro-Magnon peoples used red ocher (hematite or Fe_2O_3) in burial or fertility rites, possibly as a symbolic replacement for blood. The most dramatic early use of color was in the cave paintings of France (Lascaux and Chauvet), Spain (Altamira), and North Africa. Thus some thirty thousand years ago, humans had learned how to grind minerals, possibly mix them with a vehicle such as water, and apply them to stone surfaces as a pigment. These earth colors were relatively muted. Red was provided by red ocher, yellow by yellow ocher or sienna (a hydrated iron oxide, $Fe_2O_3 \cdot H_2O$), black from soot or charcoal (essentially pure carbon) or possibly manganese dioxide (MnO_2), white from calcined (burned) bone or a white clay such as kaolin, and brown from mixtures of hematite and manganese dioxide. Their palette contained no greens or blues and no bright colors at all.

The rise of civilizations around the Mediterranean, including Mesopotamia, Egypt, Crete, and later Greece and Rome, brought discoveries of all the missing colors. Bright yellows came from orpiment (arsenic sulfide, As_2S_3), realgar (arsenic sulfide, AsS), or lead antimonate ($PbSb_2O_5$), which had to be processed before use. Bright reds developed more slowly, but by Roman times included vermilion (cinnabar, mercury sulfide, HgS) and red lead (lead oxide, Pb_3O_4). Greens came from malachite (basic copper carbonate, $CuCO_3 \cdot Cu(OH)_2$), verdigris (a synthetic acetate of copper, $CuO \cdot 2Cu(CH_3CO_2)_2$), chrysocolla (hydrated copper silicate, $CuSiO_3 \cdot 2H_2O$), or terre verte (glauconite, a mixed silicate containing potassium, aluminum, and iron). Blues came from azurite (another basic copper carbonate, $2CuCO_3 \cdot Cu(OH)_2$), natural ultramarine (from the mineral lapis lazuli, found primarily in Badakshan in northern Afghanistan; a silicate of sodium, calcium, and aluminum), cobalt blue ($CoO \cdot Al_2O_3$), or a remarkable material, not found in nature but prepared very early by the Egyptians, called Egyptian blue (a complex mixture of copper with silicates and carbonates). New whites included chalk (calcium carbonate, $CaCO_3$), gypsum (calcium sulfate, $CaSO_4 \cdot 2H_2O$), huntite (calcium magnesium carbonate, $CaCO_3 \cdot 3MgCO_3$), and white lead (basic lead carbonate, $2PbCO_3 \cdot Pb(OH)_2$). New blacks came from galena (lead sulfide, PbS) and novel oxides of iron.

The organic world supplied the blue dye indigo, from the European herb woad or *Isatis tinctoria* and from the Asian indigo plant *Indigofera tinctoria;* the red dye madder (Turkey red) from the madder plant *Rubia tinctorum;* the red dye kermes from the female insect *Coccus ilicis* or *Kermes vermilio,* originally found in Southwest Asia; the red dye henna from the plant *Lawsonia alba* or *inermis;* the red dye lac from the Indian and Southeast Asian insect *Coccus laccae;* the yellow dye weld from the European plant *Roseda uteola;* the yellow dye safflower from the plant *Carthamus tinctorius;* the yellow dye saffron from the Asian plant *Crocus sativus* L; and the purple dye Royal or Tyrian purple from the shellfish *Murex* or *Purpuria.*

These ancient pigments and dyes may be found on a variety of artifacts, from mundane shards of pottery to precious bits of cloth, and they usually can be identified by chemical analysis. Elemental analysis can suggest an assignment—for example, by finding iron from ochers or copper from malachite. Elemental analysis is often definitive, as the presence of mercury proves vermilion, cobalt proves cobalt blue, and antimony proves lead antimonate. Diagnostic peaks in the X-ray diffraction patterns of powders can be definitive for almost all the inorganic minerals used as pigments. The analysis of atomic vibrations by means of infrared (IR) or Raman spectroscopy also can provide fingerprints of pigments. Organic dyes can be identified by their mass spectra, by their absorption (electronic) spectra in the visible and the ultraviolet regions, and by their nuclear magnetic resonance (NMR) spectra if sufficient quantities are available. Many of the methods of elemental analysis and X-ray diffraction may be made without disturbing the dye, but absorption of NMR spectra usually require displacing the dye into a solution. Although the dye remains intact, the object from which it was removed is necessarily degraded.

The group of Walter Noll, Reimer Holm, and Liborius Born established the fundamental patterns of pigments used on the earliest painted pottery. They found that single colors (monochromic decoration over the entire surface) were used first in Mesopotamia more than seven thousand years ago. The earliest such colors were black, brown, and shades of red. White came into use at about the same time that a two-color (dichromic) technique was developed. The use of greens and blues was introduced by the Egyptians some three thousand years later.

With dated pottery these scientists could track the use of a particular technique back to its geographical source. For example, by the sixth millennium B.C., potters in Iraq had learned that the natural reds of iron-bearing clays could be altered to black by heating. The process involved reduction of the iron, so that on average the ratio of oxygen to iron decreased. For example, red hematite (Fe_2O_3) could be transformed into darker magnetite ($FeO \bullet Fe_2O_3$) or hercynite ($FeO \bullet Al_2O_3$). The resulting colors ranged from rich black to dark or even pale brown, depending on the proportion of residual hematite. The black paint layer contained more iron than the pottery fabric and was sintered, or heated without truly melting. Sintering required a lower temperature than future glazes would need.

The earliest examples of black-painted pottery (sixth millennium B.C.) are called Samarra ceramics, from their Iraqi source, followed in the fifth and fourth millennia B.C. by Tell Halaf and El Obeid types (Figure 4.2). Interestingly, Obeid ceramics contain the very modern elements chromium and titanium, which must have been present in the clays. Iron-reduction black ceramics appeared in Greece and Crete during the third millennium B.C. A famous type of ceramic from Middle Minoan Crete (circa 2000 B.C.) called Kamares ware used black priming for polychrome painting. Similar materials have been found on the islands of Thera and Cyprus. By the beginning of the Late Minoan period (circa 1500 B.C.), and possibly within just a fifty-year period, the full black covering was dispensed with and true painting with red or black colors had begun. The continuation to Mycenaean, Greek Geometric, black-figured, and finally Attic red-figured pottery thus represents the evolution of a single tradition lasting five thousand years.

A parallel technology using manganese instead of iron as the source of the black color appeared in Çatal Hüyük (Figure 4.2) and other sites in southeastern Anatolia (present-day Turkey) as early as the sixth millennium B.C. From there, this technique moved into the Greek world. In this case paint slips contained high levels of manganese, although iron minerals usually were present as well. The raw material was manganese dioxide (MnO_2), but the black color contained reduced forms such as bixbyite (Mn_2O_3), hausmannite (Mn_3O_4), and so-called manganese spinels. These black phases were so stable that they could even be fired in an oxidizing atmosphere. When these black pigments were used in conjunction with hematite, products with both red and black colors could be made. Such dichromic pottery dates back at least to 5000 B.C. in Iran. Manganese black was applied to a background of red (hematite) on the clay surface, and soon white (calcite) was also used as a background.

Cinnabar or vermilion (HgS) provided a brighter red than hematite. It was being mined near Belgrade in modern Serbia by the middle of the third millennium B.C. It was used for wall paintings or decorations in Persia (550–330 B.C.), Palestine (Jericho), and numerous Roman sites. Cinnabar also was used as the coloring agent in red ink in the Dead Sea Scrolls at the beginning of the Christian era. The Romans called the pigment *minium,* and because red was the dominant color in small paintings, they became known as miniatures ("painted in red"). Manuscript titles in red became known as rubrics, from the Latin word (*ruber)* for "red."

Pigments for cosmetics were amazingly well developed in ancient Southwest Asia, even five thousand years ago. In Egypt, eyelids were painted with powdered black galena (PbS) and green malachite, lips with red ocher, and nails, palms, and soles with henna. Although much of this knowledge comes from Egyptian texts, chemical analysis of material in burials also has been critical.

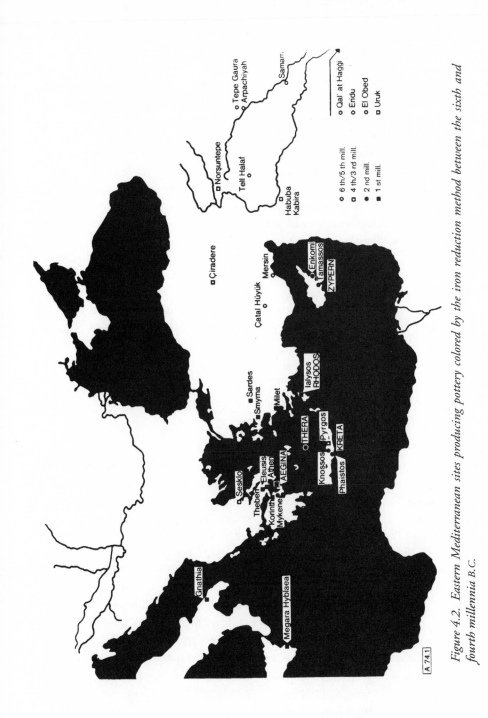

Figure 4.2. Eastern Mediterranean sites producing pottery colored by the iron reduction method between the sixth and fourth millennia B.C.

Textual records indicated that black galena eyeliner came from Asia during the Twelfth Egyptian Dynasty, from Mesopotamia during the Eighteenth Dynasty, and from Qift, Egypt, during the Nineteenth Dynasty.

The archaeological team of A. A. Hassan and F. A. Hassan discovered small piles of galena next to the wrists of a young adult female from a predynastic burial (before the first Old Kingdom dynasty) in Nagada, Upper Egypt, that dated to 3080 ± 110 B.C. by radiocarbon. Small bags that may have held the substance would have disintegrated. The presence of galena indicated that the cosmetic held a special place in the woman's life, but its abundance enabled the scientists to try to locate the geological source of the material by lead isotope analysis. Figure 4.3 shows a graphical comparison of lead-208 to lead-207, normalized to lead-206, for a variety of eastern Egyptian sources of galena, as well as for the archaeological material from Nagada. The ratios for the archaeological samples fall very close to those from Um-Anz and Zog el-Bohar, which are located some 180 kilometers east of Nagada and are accessible by the Wadi Hammamat (Figure 4.4). The somewhat nearer source of Fuakhir has quite distinct ratios. Thus this predynastic galena most likely came from a relatively nearby source and was not imported from Asia or Mesopotamia.

The Egyptian Discoveries

In contrast to the muted colors of Mesopotamian, Anatolian, and Greek pottery, those used by the ancient Egyptians were positively riotous. Plate 7 shows that the Egyptians had an essentially full palette, particularly for the production of wall paintings in tombs and temples. The illustrated example is from the tomb of Pashed, who is the small figure in the lower center kneeling by the seated Osiris. In front of Osiris, a genie is holding a pair of lighted torches, which are replicated just behind Osiris with the ujat eye. The geometrical designs on the left are the stylized representation of a desert. Although the pigments of this wall painting have not been analyzed, the reds, browns, and dark yellows probably are hematite; the green of the faces and of the bird's wing, malachite; the yellow of the bird's legs and feet, orpiment; the black surfaces, carbon; the white surfaces, gypsum or calcium carbonate; and the deep blue of the genie's headdress on the right, Egyptian blue.

Chemical analysis continues to add to our knowledge of the Egyptian palette. H. Riederer used X-ray diffraction and infrared spectroscopy to identify white huntite on terra-cotta Nubian bowls from about 1600 B.C., and on the lid of a terra-cotta canopic jar from an eighteenth-century B.C. burial. The lid is in the shape of a man's head, painted bright white with huntite and highlighted with carbon black and yellow orpiment. Although the synthetic Egyptian blue

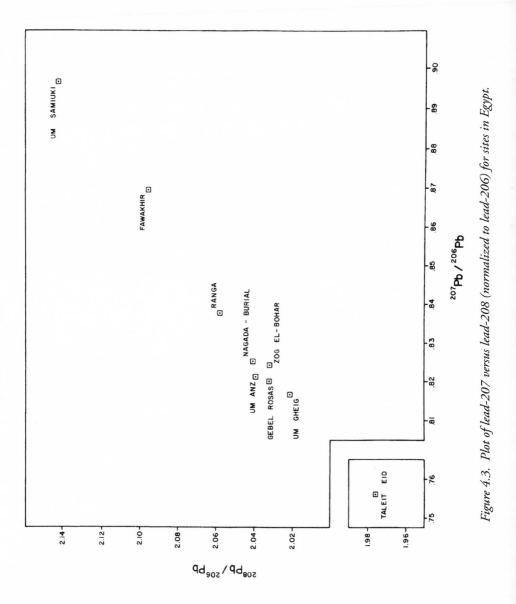

Figure 4.3. Plot of lead-207 versus lead-208 (normalized to lead-206) for sites in Egypt.

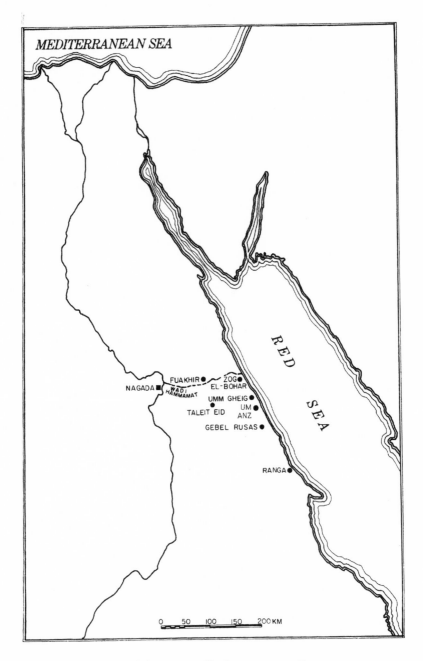

Figure 4.4. Sources of lead ore in eastern Egypt.

pigment provided the normal source of blue, Riederer found that two terra-cotta vases from the Amarna period (about 1370 B.C.) were decorated with red and violet geometric outlines filled in with cobalt blue.

This blue pigment has been controversial in such an early context. Unlike lead, cobalt has no isotopic signature that permits the determination of provenance. A. Kaczmarczyk used the pattern of elements associated with cobalt to uncover the source of Egyptian cobalt. His study centered on the blue glass-like material composed of glazed quartz, that is often referred to by the misnomer Egyptian faience. Along with cobalt, he always found elevated amounts of manganese, iron, nickel, and zinc in glazed quartz from the New Kingdom (sixteenth to eleventh century B.C.). Kaczmarczyk found the same elemental pattern in alum deposits in the North African Dakhla Oasis, leading him to conclude that Egyptian cobalt came from this source. Pink or lavender veins in the alum sometimes contained as much as 3.5 percent of cobalt. He also discovered that the few Mycenaean samples of cobalt blue glass (fourteenth to thirteenth century B.C.), which are contemporaneous to the New Kingdom, had this same pattern of metals, leading him to conclude that this cobalt also had come from the Egyptian oases.

The earliest known use of cobalt in glass as a pigment is a piece from Eridu, Iraq, that has been dated to the twenty-first century B.C. This and later Mesopotamian and Syrian cobalt glasses do not show any correlation with manganese, iron, or zinc, which would have indicated an Egyptian source. The presence of arsenic and the absence of manganese correlate with the characteristics of well-known sources of cobalt in Persia. These same sources much later provided cobalt for classic Chinese porcelain.

By far the most remarkable chemically of the new pigments is Egyptian blue, which was not known at the time to occur in nature. It had to be prepared by a complex formula that required careful selection, preparation, and blending of materials, and control of temperature in a firing process. Egyptian blue was created sometime before 2500 B.C., at least by the Fourth Dynasty. It found widespread use in Southwest Asia, later spread to the Greek and Roman worlds, and was lost to civilization by A.D. 400. The blues of Pompeii paintings, for example, are from Egyptian blue. Its formulation was described in detail by the first-century writer Vitruvius, but was not recreated until the nineteenth century. Numerous cakes of Egyptian blue frit have been discovered. A *frit* is a glassy substance used as a raw material for making glass or glaze. The chemical composition of Egyptian blue is quite distinct from that of Egyptian glass or glazed quartz.

The Egyptian formula, as clarified by the nineteenth-century workers and more recently by Thomas Chase, Michael Tite, and co-workers, consisted of sand (silica, silicon dioxide, SiO_2), calcium carbonate, a source of copper (either malachite or raw copper), and possibly sodium salts to serve as a flux to reduce the temperature of firing. The overall chemical formula of the product as a pure material approximates to $CaCuSi_4O_{10}$ ($CuO•CaO•4SiO_2$), which corresponds to the later-discovered mineral cuprorivaite. Chase found that firing at about 850°C suffices and that much higher temperatures destroy the substance. Copper also can be converted to a red form that sometimes is apparent as an impurity.

Tite and co-workers carried out extensive elemental analyses of Egyptian blue samples not only from Egypt but also from the Southwest Asia sites of Nimrud and Nineveh and sites in western Europe from the Roman period. Figure 4.5 uses a triangular graph to display the relative amounts of silica (SiO_2), calcium oxide (CaO), and copper oxide (CuO) in these samples. The

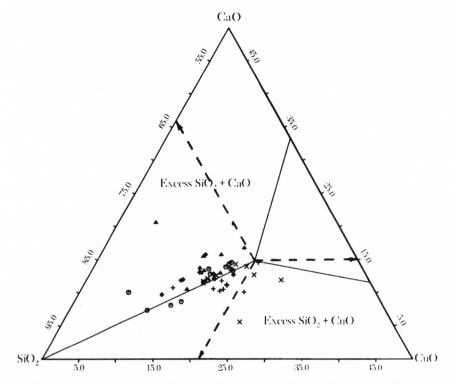

Figure 4.5. Ternary plot of silicon dioxide (SiO_2), calcium oxide (CaO), and copper oxide (CuO) in samples of Egyptian blue. The point at which the three solid lines converge represents the composition of the natural mineral that corresponds to Egyptian blue. The three dashed lines indicate how to obtain the elemental levels for the mineral. Key: ⊙ Egypt, ▲ Amarna, + Nimrud, x Nineveh, ◆ Roman. Reprinted with permission. © 1984 by the American Chemical Society.

respective levels of SiO_2, CaO, and CuO are found on the left, right, and bottom sides of the triangle. Any point within the triangle corresponds to a substance made of these three components, whose amounts are found by drawing three lines from the point to the axes. The lines for a particular component are drawn parallel to the side of the triangle opposite the apex representing 100 percent of that component. The point at which the three solid lines come together represents the formula of the mineral cuprorivaite (about 22 percent CuO, 15 percent CaO, and 63 percent SiO_2—obtained by following the dashed lines).

The observed compositions lie roughly along the line from the ideal cuprorivaite point to the SiO_2 apex on the left, which represents pure silica. Thus the samples of Egyptian blue invariably had more silica than in the mineral cuprorivaite. The amount of calcium oxide in the samples fell mostly in the range 8–17 percent, a much larger proportion than in Egyptian glass (5–10 percent). Moreover, Egyptian glass, as will be seen in Chapter 5, usually also had 10–20 percent sodium oxide, whereas Egyptian blue rarely had more than 4 percent. (Sodium oxide is not shown in the triangular plot of Figure 4.5.) As raw material for Egyptian blue, the Egyptians may have exploited a particular type of sand that was calcium rich and sodium poor.

Mediterranean Dyes

For fabrics, bright colors had to await the development of organic dyes and of mordanting and vatting techniques to assure color fastness. By some point in the second millennium B.C., fabric dyeing was a big business in the eastern Mediterranean, and the most famous of those dyes was Tyrian or Royal purple. This material required complex chemical processing through vatting of secretions from several molluscan species, including *Murex trunculus, Murex brandaris,* and *Purpura haemastoma* in the Mediterranean (Figure 4.6), and other species exploited later in China, Japan, Peru, and Mesoamerica. The earliest and strongest associations are with Levantine (eastern Mediterranean) sites such as Tyre or Sarepta (present-day Sarafand). These cities were part of the Canaanite or Phoenician world, and the names of both these peoples have been alleged to derive from etymological roots meaning "purple."

The production was enormous, as it took more than five thousand mollusks to produce only one gram of the dye. Shell middens of considerable extent grew up not only at Tyre and other Levantine sites such as Sidon, but also on Crete. Akkadian and Ugaritic texts from the Late Bronze Age in Southwest Asia mention the dye, and biblical references occur as early as Exodus (26:1). Legend associated the dye with Hercules. Its importance was recognized both by the Roman scholar Pliny the Elder, who gave a detailed and generally accurate description of its production, and by the Roman emperor Nero, who issued a decree restricting its use on garments to himself. Even in Roman Republican days only two censors and generals during a triumph could wear an entirely purple toga; consuls and praetors had to be satisfied just with purple edges to their garments. Cleopatra, however, managed to produce an entirely purple sail in her attempt to support Antony in the sea battle of Actium.

The manufacturing process from live mollusks not only produced the incredible stench described by Mailer in *Ancient Evenings,* but also restricted dye production to coastal regions. In the fourth century A.D., however, the discovery of a method for production from dead mollusks permitted dyehouses to move inland. The destruction of the western Roman Empire terminated its production in most of Europe, but the Byzantine Empire continued its use. The Byzantine writer and royal daughter Anna Comnena emphasized that royal sons were to be born in a particular purple-colored room in the palace, so that they would be truly *porphyriogenatos,* "born in the purple." The fall of Constantinople in 1453 brought an end to the use of Tyrian purple, except in some local cultures.

Modern chemical analysis has demonstrated that the purple ingredient of the snail is a molecule called dibromoindigotin, very closely related to the blue dye indigotin (or indigo, from its original plant source in India). The structures of these molecules are shown in Figure 4.7, and they differ only by the presence of two bromine atoms in the purple dye. The illustrated structures use the cryptic symbolism of organic chemistry, whereby carbon and hydro-

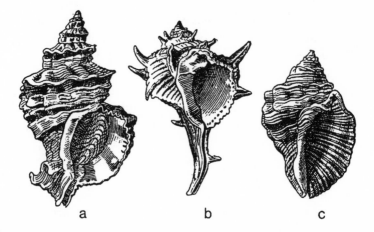

a   b   c

Figure 4.6. Shells of the principal Mediterranean mollusks that yield dibromo-indigotin and related dyes: a, M. trunculus; b, M. brandaris; c, P. haemastoma. *Reprinted with permission.* © *1990 by the American Chemical Society.*

gen atoms are understood but not depicted. Carbons are present at any angular vertex or at the end of a dash, but the usual letter C is not written. Each carbon must have four bonds, represented by short lines or dashes. If a carbon center lacks all four lines, unexpressed hydrogen atoms fill out the quartet. Thus the symbol = represents $CH_2=CH_2$, the molecule ethylene. Note that all three molecules in Figure 4.7 have two carbonyl groups (C=O), but the carbon is undepicted.

Production of Tyrian purple involves reduction of the natural product to a water-soluble leuco form that may be absorbed by the fabric and then reoxidized to the colored form in air. This process explains what the suffix *tin* is doing at the end of the word. Tin is an effective reducing agent for the conversion of the natural product to its leuco form. Different mollusks produce varying amounts of the purple dibromoindigotin and the blue indigotin. Whereas *M. brandaris* and *P. haemastoma* produce mainly the dibromo form, *M. trunculus* provides both materials. Choice of a mix of species can thus control specific shades of color. A process involving an initial *Murex* bath followed by a *Purpura* bath was called double dyeing and is mentioned frequently in the Bible.

Huge piles of shells and literary references are not the only vestiges of this important industry. Archaeologists have actually found samples of the dye itself and have been able to analyze it. Patrick McGovern, with his colleague Rudolph Michel, was able to examine a purple sediment from thirteenth-century B.C. jars found at a dyeing and pottery factory at Sarepta, on the coast of Lebanon. The fact that the color remained after 3,300 years certainly attests

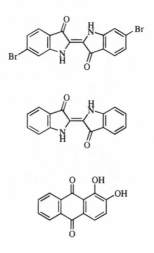

Figure 4.7. The chemical structures of, top, dibromoindigotin; middle, indigotin; and bottom, alizarin.

to its fastness. The shape of the pot (Figure 4.8), with its wide mount for introduction of molluscan raw material and a spout at the bottom for draining off the dye solution from the residue of the crushed animals, certainly supports its use in dye production. McGovern and Michel compared the infrared spectrum of the purple residue with that of modern dibromoindigotin and found a satisfactory coincidence. They also proved bromine by elemental analysis. This study thus documented the earliest example of Tyrian purple by direct archaeological chemistry.

The blue form (indigotin) was available more readily from plant sources, woad and indigo. Woad, for example, was the dye used by the Picts (from the Latin *pictus*, or "painted") for body painting in northern Scotland after the third century A.D. Biblical blues could have come from either the nearby molluscan sources or from these plants. In Numbers (15:38) the Lord enjoins the children of Israel to wear garments with fringes of white and of the color *tekhelet*. The meaning of the Hebrew word *tekhelet* is unknown but is considered to be blue because, as stated by Rabbi Meir, it resembles the sea, which resembles the sky, which resembles sapphire, which resembles the "Throne of God." One use of the color, as described by the Talmud, was to determine when to recite the Sh'ma (Creed) in the morning, namely as soon as tekhelet could be distinguished from white on the garment fringes. Because the formula for tekhelet was lost by A.D. 760, orthodox prayer shawls have had color-

Figure 4.8. A pot from Sarepta, Lebanon, presumed to be used in the production of Tyrian purple. The spout at the bottom may have been used to drain off the dye solution from the residue of crushed mollusks.

less tassels, since an incorrectly formulated dye would be impious. Tekhelet probably is indigotin or a mixture of indigotin and dibromoindigotin.

The ancient red mordant dye madder was second to Tyrian purple in fame and importance. Fabrics containing madder have been found both at Mohenjo-Daro on the Indus River and in Egyptian tombs. It is made fast by the use of the mordant alum (hydrated potassium aluminum sulfate, $KAl(SO_4)_2 \cdot 12H_2O$). The manufacturing process involved numerous complex steps. Although madder was native to Asia, it was introduced to Europe during the Crusades. Madder dyeing was important right up to the Industrial Revolution and may, for example, have provided the color for British fox hunters and military uniforms (Redcoats, to the Yankees). In one of its later manifestations it was called Turkey red. The madder root contains several related colored materials, but alizarin, whose organic formula is also given in Figure 4.7, is the most important. The two hydroxy groups (OH on the right ring) provide a means of attaching the dye to the oxygen atoms of the mordant alum.

The dye madder can be converted to a pigment by embedding it in a lake. Madder grows naturally on Cyprus and has long been a popular Easter egg dye there, as well as in Greece, Russia, and the Ukraine. G. V. Foster and P. J. Moran found that madder was used as a surface colorant on Cypriot pottery from the eighth and seventh centuries B.C. Like Tyrian purple, the color proved to be fast over thousands of years, and archaeologists identified it initially by visual examination. Its presence was proved by extraction from a chip of pottery and examination by mass spectrometry, which provides the relative weights of molecular pieces. In the chemist's counting system, the atom hydrogen weighs about one atomic unit, and larger atoms are measured in approximate multiples of the weight of hydrogen (carbon is 12, oxygen is 16). The total weight or mass of a molecule, called its molecular weight, is obtained by adding the weight of each atom in the molecule. Figure 4.9 shows the mass spectrum of a sample of a purple-red dye extracted from a Cypriot pot. The largest peak at the high mass (right) end is 240, which in fact is the molecular weight of alizarin. The other peaks arise from other molecules and from fragments of the alizarin molecule. Mass spectrometry is so sensitive that the experiment may be carried out on fractions of a milligram. Alternatively, the presence of madder may be proved by comparing its infrared spectrum with that of authentic material.

Alum was not available on Cyprus until the Roman period, so another mordant had to be used for these Cypriot Geometric and Archaic pots. The question of the identity of the mordant is common to madder coloring found on pots and even marble or limestone figures from the twelfth to the eighth centuries B.C. in Israel and Greece. Madder was used in ancient Egypt on limestone, wood, and plaster but not on pottery. Elemental analysis by X-ray techniques on the Cypriot pots showed levels of iron in the pigment that were much higher than in the body of the pottery. Foster and Moran suggested that iron served as the mordant to bind madder to the matrix of a lake, which was powdered and used in the usual fashion as a pigment.

The active ingredient of the insect dye kermes is similar to alizarin and may be distinguished easily by mass spectrometry or infrared spectroscopy. *Kermes* comes from the Armenian word for "worm" and is the root of our word *crimson*. Kermes was the most important Babylonian red dye and was used by the Egyptians to dye fabrics.

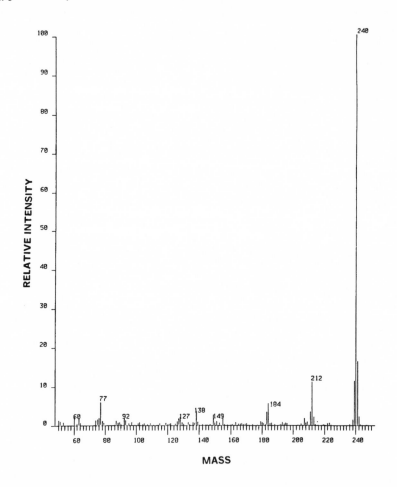

Figure 4.9. Mass spectrum of the extract from a purple-red chip from a Cypriot pot. The highest principal peak has a mass of 240 and corresponds to the molecular weight of alizarin.

It is interesting that the classical Greeks and Romans introduced very few novel pigments or dyes, two exceptions being white lead and verdigris. They depended on imported Egyptian blue. Cobalt blues and copper reds became more common. Although copper usually conveys a blue or green color, the less usual oxidation state with one positive charge on copper (cuprous oxide) or elemental copper causes the color known as copper red, found not only in the classical world but also in Egypt and Iron Age Britain. Copper red is much brighter than iron red but sometimes appeared in unwanted contexts. The copper in Egyptian blue could be converted inadvertently to copper red by incorrect firing conditions, and in some cases it was even painted over with another blue.

Relics of the classical period include written descriptions and bulk samples of both dyes and pigments. Theophrastus, Vitruvius, and particularly Pliny the Elder in his *Natural History* gave detailed descriptions of the sources of raw materials and the procedures for the preparation of numerous pigments, including Egyptian blue, verdigris, and lead white. Archaeologists have found dye lakes and bulk pigments, the most dramatic of which may have been from Pompeii and Herculaneum. Pigments in jars and bottles had been protected from light and sealed off from air since the eruption in A.D. 79 and hence were examined in very nearly their original condition.

Dyes and Pigments of the New World

New World pigments and dyes rivaled those of the Old World in variety of color and probably in technology of production as well. The expertise of New World artisans has been revealed most dramatically in ancient Mesoamerican frescoes and South American textiles. The frescoes of Bonampak in the jungles of Chiapas, Mexico, discovered in 1946, provide an almost unaltered example of the scope of Mayan art (Plate 8). The colors include not only black, red, yellow, brown, and white, but also blue and green—a full palette. Their construction was dated to about A.D. 785, from a contemporaneous stele (a carved statuelike stone). An older fresco at Uaxactún in Guatemala, dating to before A.D. 633, shows that the expertise was widespread.

All the Mesoamerican pigments but the blue have sources similar to those of analogous colors in the Old World. The blues of the Old World came from either copper or cobalt minerals, but the Maya blue was based on combining the organic vegetable dye indigo, identical to the material whose structure is depicted in Figure 4.7, with a white clay, called attapulgite or palygorskite, to form a lake. The fastness of Maya blue is truly impressive, as it resists the effects of dilute acids and bases, solvents, oxidizing and reducing agents, normal heat, and natural biological corrosion. The usually sensitive organic molecules are protected by being encapsulated within the inorganic clay structure.

H. Van Olphen first suggested the combined organic/inorganic composition in 1966, but for some time no deposits of attapulgite were found. A major mining source finally was located in a cenote (a natural limestone sinkhole or well) in the town of Sacalum, near the Mayan ruins of Uxmal. Mineralogical analysis showed that the site contained vast quantities of pure attapulgite, and with some chagrin it was recognized that *Sacalum,* in the local Mayan tongue, corresponds to *sak luʾum* and means "white earth." D. E. Arnold and B. F. Bonor calculated that between 300 and 600 cubic meters of clay had been removed from the mine, whose access was from the bottom of the cenote. They have suggested that this site was the principal source for the clay used for Maya blue, although its existence is recorded historically only back to 1549. Pottery from the period A.D. 800–1000, however, was found in the cenote. The use of Maya blue extends well beyond the Yucatán Peninsula. Several objects painted with Maya blue, for example, have been found at Zaachilá in the Mexican state of Oaxaca.

New World artisans also developed a red dye that proved to be superior to existing Old World dyes. Cochineal had been produced since 1000 B.C. or earlier from the body of the insect *Dactylopius coccus,* a small worm related to the louse that lives on prickly pear cacti. Prickly pear *(Opuntia)* could be grown and the insects harvested almost as if they were domesticated. Although it took about two thousand female insects to produce a gram of dye, the cochineal insects could be harvested several times a year compared with only once a year for the Old World kermes insect. Moreover, cochineal has ten times more of the active coloring principle than does kermes. Their respective coloring agents, called carminic acid and kermesic acid, are related to alizarin (Figure 4.7).

The Spanish arrived in Mexico in 1518, and by 1550 the dye was already being widely exported to Europe and quickly superseded kermes. For over two centuries cochineal was produced in Mexico for the European market, primarily by indigenous farmers, until Spain lost its Latin American possessions in the early 1800s. Production was moved to the Canary Islands and continued until synthetic aniline dyes replaced it later in the nineteenth century. The major use of both kermes and cochineal, mordanted with aluminum, tin, or iron, was in dyeing cotton; madder has never been successful with cotton. Plate 6 shows the result of yarn dyed with cochineal that had been mordanted with different metals. Both Joshua Reynolds and J. M. W. Turner experimented with cochineal as a pigment to produce lively flesh tones or fiery landscapes, but unfortunately these colors have faded to brown.

The textiles of South America, particularly those of Peru, have not suffered the same fate. Many of them, found as wrappings for burials in desert climates, have maintained their bright colors for hundreds of years. The mantle illustrated in Plate 9 was made around 100 B.C. from alpaca fibers. Analysis of the dyes, however, proved to be complex, as the reds could have come not only from cochineal, but also from plant sources such as relbunium *(Relbunium ciliatum)*, which is related to madder. Madder and relbunium contain alizarin, purpurin, and other coloring agents, which differ only in proportions. The distinction among madder, relbunium, and cochineal, as well as indigo (available either from *Indigofera suffructicosa* or *Cybistax antisyphylitica* to Peruvians) can be made from absorption patterns in the ultraviolet and visible parts of the electromagnetic spectrum. (The ultraviolet region is 200–400 nm.) This method was developed by Max Saltzmann. Kathryn Jakes used infrared as well as visible spectroscopy to identify the dyes. Figure 4.10 shows how she identified a hair fiber from Paracas, Peru as being dyed with relbunium, from the characteristic double maximum in the visible spectrum that is different from the (unillustrated) cochineal pattern.

The pre-Columbian civilizations in the New World also had great artisans who worked with metal. They could create a gold surface from a baser alloy through a process known as depletion gilding (Chapter 7). Use of an alloy of about 25 percent gold with copper and possibly silver, called *tumbaga*, was more economical than use of pure gold. Treatment with acidic materials dissolved copper and silver from the surface and left the gold in place, giving the object an appearance of being entirely of gold. Such a process is related to the European decorative etching of armor or swords, as discussed in Chapter 7, and eventually to printer's plates.

Dyes and Pigments of East Asia

The Asian world provided many innovations in the use of dyes and pigments, although it did not create many novel substances that found widespread use. Certainly one of the exceptions to this generalization is Chinese ink, known to much of the Western world as india ink. Early literacy among segments of the Chinese population provided a need for an inexpensive, readily available writing material. China ink probably dates to the Wei Dynasty (A.D. 220–265) or earlier, and by the fifth century its manufacture had become relatively standardized. It is composed of a mixture of soot or lampblack with glue as a binder, just like other paints. As these materials are normally immiscible, the ink is not a solution but a suspension of the finely divided hydrophobic (water-insoluble) carbon with the hydrophilic (water-soluble) glue. The carbon pigment originally was soot produced by the controlled burning of pine wood, but by the eleventh or twelfth century it was lampblack produced by burning oil with a wick. The glue was produced from animal sources such as deer antlers, animal skins, or fish. The actual manufacture involved grinding and sieving the

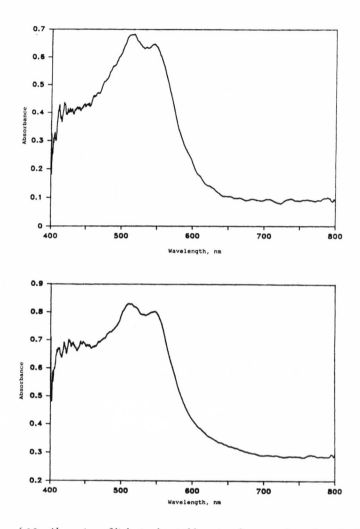

Figure 4.10. Absorption of light in the visible region for, top, *alpaca hair dyed with an authentic sample of aluminum-mordanted relbunium, and,* bottom, *a sample of colored hair fiber from Paracas, Peru, funeral bundle 421.*

carbon, mixing it with the glue, and then extensively kneading, pounding, steaming, and rolling to produce an intimate mixture of the components. The technique was imported to Korea and Japan by the Tang Dynasty, which began in A.D. 618, and before long was ubiquitous in Asia for writing and painting.

Painted pottery appeared in China by 3000 B.C. Before the end of the second millennium B.C., during the Shang Dynasty, glazes made from wood ash and clay had become common. The addition of coloring agents evolved through several steps to the cool olive-green ceramics known as celadon, which attained a bluish tint during the Southern Song Dynasty and also were widely produced in Korea. The opalescent jade color of celadon required the presence of iron, but was brought out in the feldspathic glaze (from powdered granite) by multiple reflections caused by the presence of a multitude of tiny bubbles. The bubbles, presumably from outgassing of water, scatter light coming from any angle.

The Chinese invention of hard-paste porcelain inspired remarkable colorations in addition to the well-known blue and white (Plate 5). The lustrous red of sang-de-boeuf ("ox blood") porcelain (Plate 10) comes from the presence of copper in the glaze. Porcelain blue is attributed to the introduction, from Persia during the fourteenth century, of cobalt pigments that lacked the manganese impurity that had dulled earlier Chinese blue glazes. The use of cobalt as a deep blue pigment has been discovered several times. We have already seen its use in Iraq by 2100 B.C., in Egypt by 1370 B.C. in wall paintings, and by the Mycenaean, Mesopotamian, and Syrian cultures of the first and second millennia B.C. The tradition continued or was rediscovered in the Islamic world. Tin-glazed pottery (majolica or faience, to Europeans) had been discovered in Samarra on the Tigris River during the Abbasid Caliphate (ninth century A.D.). The glaze was produced in both white and blue varieties, and the blue color was produced by the intentional addition of a cobalt blue pigment. This inglaze pigment is identical to the underglaze blue of Persian faiences from the twelfth to the fourteenth centuries. It was this Persian blue pigment that was recognized and then exploited by the Chinese for their porcelain. For both periods the mining source for the cobalt pigments may have been the village of Qamsar (or Khamsar) in central Iran, which is mentioned in the treatise of Abu'l Qasim from 1302.

Painted pottery in Japan dates from about the third century A.D. Complex wall paintings were produced in the Fukuoka and Nara regions during the fifth through seventh centuries, particularly after the introduction of Buddhism. Analysis of these materials has indicated that the red was from red ocher and vermilion; yellow, from yellow ocher; green, from malachite; blue, from azurite; gold, from powdered gold; white, from white lead or kaolin clay; and black, from China ink. Many authentic samples of these pigments dating from the eighth century may be found in the Shosoin repository in Nara.

To the Western eye, Japanese art is epitomized by the prints on paper or silk that have been called *ukiyo-e* ("pictures of the floating [or sad] world"). This genre includes the familiar fingerlike ocean waves and views of Mount Fuji in the works of Katsushika Hokusai from the early nineteenth century. In contrast to the Western tradition of using linseed oil as a binder, which hardens through polymerization and evaporation, the glue for *ukiyo-e* gelatinizes as the water vehicle evaporates.

The *ukiyo-e* palette included the same pigments as had been used hundreds of years earlier for wall paintings, although the white may have come from oyster shells (calcium carbonate) rather than clay. The pigment was mixed into a paste made of water and glue. The paper or silk could also be colored with organic dyes, which included red from safflower or cochineal (from China, not America), blue from indigo or dayflower (and later Prussian blue), yellow from gamboge resins from *Garcinia* evergreens, browns and blacks from tannins, and greens from mixtures of indigo and yellow dyes. The fugitive nature of these colors and the porosity of the pigment layers make this artwork particularly susceptible to deterioration.

Contributions from Europe

In Europe the loss of social order and the rise of Christian and Islamic institutions followed the decline of the Roman Empire. Monastic orders not only preserved classical texts but also carried out empirical scientific research, much of which we might now call alchemy. The direct synthesis of cinnabar (vermilion) by the combination of elemental mercury and sulfur, by the tenth century, was a result of the monastic and Moorish willingness to experiment. Art again became a layman's pastime in the twelfth century, particularly in Lombardy, Italy. By the fourteenth century, independent apothecaries began to serve as color merchants. One of the problem colors always had been blue. The long tradition of Egyptian blue had been lost with the fall of Byzantium. The copper-based blue of azurite was not a deep color, and cobalt blues and ultramarine were never widely known. Sometime about 1475, possibly in Saxony, the pigment smalt was invented. A cobalt ore called saffre was fused with silica and potash at a high temperature to produce a glass that was ground up to form the pigment. But not only was it not an intense blue, it tended to decompose in the presence of linseed oil, the binder for most European painters. In 1704 a new blue pigment was produced by the accidental combination of iron salts with cyanide. The precipitate was expected to be white but turned out to be blue from the presence of hydrated ferriferrocyanide, $Fe_4[Fe(CN)_6]_3 \cdot nH_2O$ ($n = 14$–16). Known as Prussian blue, this pigment is still in use, although it has a tendency to fade.

Knowledge of the structure of these new pigments permits us to identify later repairs and even to establish that a particular piece is fraudulent. In their survey of artwork in Byzantine manuscripts from the tenth to the thirteenth centuries, Mary Virginia Orna and her collaborators found common pigments such as azurite, ultramarine, vermilion, and white lead, as well as a surprising number of organic dyes, but in one case they found the clear fingerprint of Prussian blue. This manuscript from the University of Chicago, called Archaic Mark, had an unclear history but was suspected to be a nine-teenth-century version of the Greek gospels. The cyanide part of the Prussian blue molecule contains a carbon-nitrogen triple bond ($C{\equiv}N$), which unmis-takably absorbs in the infrared at about 2080 cm^{-1}. Figure 4.11 compares the infrared spectrum of a sample of the blue pigment from Archaic Mark with that of an authentic sample of Prussian blue, and the peak at 2080 cm^{-1} is unmistakable. The remaining peaks in the manuscript sample may be attrib-uted to the organic binding agent. The manuscript thus is probably a recent copy, since Prussian blue was not discovered until 1704.

Illustration or illumination of medieval Bibles involved the intricate use of pigments, and their analysis requires the subtle characterization of minute particles. A group led by Robin J. H. Clark has used Raman spectroscopy, closely related to infrared, to identify pigments. Most of these pigments are seen in the single letter I (Figure 4.12), from the initial word in the book of Genesis ("*In principio,*" "In the beginning") in a Bible dating to circa 1270 called the Lucka Bible, from its long stay at the Abbey of Lucka in Znojmo in the Czech Republic. The letter is 83 millimeters high and contains depictions of the seven days of creation. The alternating yellow and orange frames proved to be orpiment and red lead, respectively. The blue backgrounds of

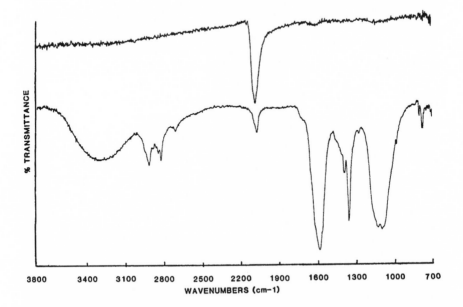

Figure 4.11. Infrared spectra of, top, *Prussian blue and,* bottom, *a blue pigment from the Archaic Mark manuscript. Reprinted with permission.* © *1989 by the American Chemical Society.*

Figure 4.12. The letter I from the first word in the Book of Genesis of the Lucka Bible.

four of the scenes are lapis lazuli or natural ultramarine. God's tunic, in the fourth and seventh scenes, is pure red lead, but the red dye in the adjacent text is vermilion.

One of the most difficult colors to represent is gold. Pigments of this color on canvas or walls in fact usually come from suspensions of gold particles. Attachment to fabrics was an even greater challenge. (Recall the legend of the Golden Fleece.) The first solution was to use flat gold strips or to wind very thin strips of gold around silk yarn, but the result was not at all flexible. A superior material, called membrane thread, appeared in Asia in the eleventh century and was introduced to Europe by the fourteenth century. By this procedure, powdered gold was mixed with a binder such as egg white and applied to leather, animal gut, parchment, or even paper. Narrow strips of these gilded materials then were wrapped laboriously around a fibrous core to form the thread. Figure 4.13 shows a highly magnified comparison of both solid gold wound around a thread and the membrane thread. The new material not only was more flexible than metal-wound strips, but it also weighed much less and hence made weaving a lot easier. By 1600, however, membrane thread technology was replaced by the use of gilded silver threads.

Possibly the most famous artifact from medieval Europe is the Shroud of Turin. Although alleged to date to the time of Jesus, the radiocarbon date proved to correspond to about A.D. 1325, close to its first historical record. The date aside, it is still unknown how the striking image of a crucified man was applied to the cloth. The 1978 examination of the fabric found no trace of a pigment and concluded that the yellow image was inherent in the fabric. Walter

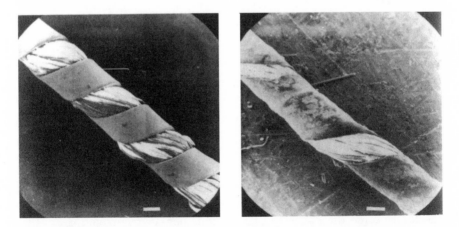

Figure 4.13. Left, *solid gold metal strips wound around yarn.* Right, *membrane thread (gilded strips of leather or parchment wrapped around a fibrous core). The white scale bar represents 0.1 mm.*

McCrone, however, found particles of red ocher, which he suggested was the missing pigment. The final word is not in.

The invention of movable type by Gutenberg represents the beginning of the information revolution that continues today. In order to accomplish this invention, Gutenberg not only had to develop the metallurgy associated with making the type itself and conceive of the mechanics of the process, but he also had to invent or procure an ink that would be compatible with it. The chemical analysis of the paper and ink on Gutenberg Bibles must be fully nondestructive because of the value and importance of the objects. The method chosen for elemental analysis of the ink is called proton- (or particle-) induced X-ray emission (PIXE), which, besides having one of the finest acronyms in analytical chemistry, can determine the levels of a wide range of elements nondestructively on a spot smaller than the width of a letter. In addition to carbon-based inks such as China ink, there also are iron-gall inks, composed of crushed oak-nut galls and copperas (hydrated iron sulfate, $FeSO_4 \cdot 7H_2O$, also called green vitriol), with gum arabic as the binder. Copperas forms a very dark solution with the tannins in the gall. Such inks would be identified by high levels of iron.

PIXE analysis found only copper and lead in Gutenberg's ink, no iron. Although the technique was not sensitive to the presence of carbon, the lack of iron is consistent with a carbon ink. Gutenberg is thought to have used six different batches of ink. The copper-to-lead ratio distinguished the six batches roughly (Figure 4.14). The elements present in the paper were subtracted out by computer through analysis of unpigmented paper. The ratio of lead (Pb) to calcium (Ca) is a good measure of the amount of ink, because it includes both a measure of the ink (Pb) and of the paper (Ca). Not only does the plot distinguish the various inks used in the Gutenberg Bibles, but it also succeeds in differentiating inks in other pre-1501 books, which as a group are termed *incunabula*. The positions on the plot for the ink in the thirty-six-line Bible and for works of Aquinas printed by Gutenberg's associate Schoeffer suggest that the former may have been printed by Gutenberg and that Schoeffer may have worked with a similar but clearly distinct ink formula than that of Gutenberg.

By the time of the Enlightenment, artists not only had a rich palette (it is said that Tintoretto worked with four different blue pigments), but had also mastered the process of blending and overlaying pigments, which required knowledge of the chemical compatibility of the materials. With the Industrial Revolution came the discovery of numerous new pigments and dyes. Two excellent blue dyes appeared early in the nineteenth century. During the chemical manufacture of soda, which involves heating sodium sulfate with coal and limestone, a blue encrustation was observed within the furnace. Exploitation of this observation led to what was called synthetic ultramarine, whose complex structure is composed of sodium aluminosilicates with polysulfide groups trapped within the crystal. In 1802 Louis J. Thénard produced modern cobalt blue (cobalt aluminate, $CoO \cdot Al_2O_3$), which eclipsed smalt within a few years and became a favorite of Impressionists for depicting sky. An entire family of pigments grew up around the element chromium: chrome yellow ($PbCrO_4$), chrome red and orange ($PbCrO_4 \cdot PbO$, with the hue dependent on the ratio of the two components and on the particle size), and chrome green ($PbCrO_4$ mixed with Prussian blue). The cadmium family of reds, oranges, and yellows appeared in the twentieth century, along with titanium white (titanium oxide, TiO_2) (1916), which is the most important white pigment and opacifying (covering) factor in current paints. In 1935 an entirely new type of pigment was created that included not only metals but also organic structures. The intense phthalocyanine blue ($CuC_{32}H_{16}N_8$) is a representative of what are now called organometallic compounds, and in fact is related to the red coloring agent in blood, an organometallic called hemoglobin.

The discovery of the synthetic dye mauve forever altered the way chemists approached the development of dyes. In 1856 William Henry Perkin (called "Perkin, Sr." by chemists to distinguish him from his son of the same name,

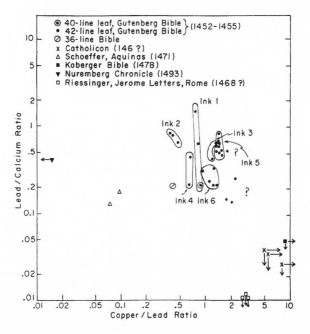

Figure 4.14. Plot of the copper/lead ratio versus the lead/calcium ratio in incunabula inks.

who at the time of his death in 1929 was one of the most eminent chemists of his generation) was trying to synthesize quinine for the treatment of malaria. (The successful synthesis did not occur until 1944.) His mixture of coal tars at one point yielded a beautiful purple material ("aniline purple") that he was able to isolate. He developed a method of manufacture, christened the dye mauve or mauvein, and became rich and famous. At the time of the discovery, Perkin was the eighteen-year-old assistant of August Wilhelm Hofmann. Research on coal tars in several countries yielded numerous other dyes. Hofmann left London in 1865 for Berlin, where he helped found Germany's dye industry with the discovery of aniline red and aniline blue. A French group under Verguin discovered yet another aniline dye, first called fuchsin and later magenta, after a French military victory over the Austrians in the Italian city of that name.

Within thirteen years of Perkin's discovery, aniline or coal tar dyes had almost entirely supplanted natural dyes, producing brighter colors and a much wider range. These dyes, however, were unfortunately fugitive. In 1869 both Perkin and Heinrich Caro succeeded in the practical synthesis of alizarin (Figure 4.7), the active ingredient of madder, which was known to be colorfast. Its production was instrumental in the growth of the giant German dye manufacturer, BASF (for Badische Anilin und Soda Fabrik). It was chemists at BASF who in 1897 succeeded in developing a practical synthesis of indigo, and within a few years the production of natural indigo had been devastated. The wide availability of synthetic indigo enhanced the popularity of dyed denim (from the French town of Nîmes) or blue jeans (either from the French town Gènes or the Italian town Genoa).

The human perception of color has driven chemical creativity for thirty thousand years or more. Our ancestors began by selecting attractive minerals and figuring out how to make them adhere to a surface such as a cave wall. They learned how to modify the color by oxidation state selection through heating. By 2000 B.C., humans had begun extracting colored materials from plants and animals. These substances often required significant chemical modification with mordants or in vats to enable them to adhere permanently to fabrics or other surfaces. In Southwest Asia or Egypt during the third millennium B.C., artisans first succeeded in creating a major new chemical substance, Egyptian blue, by combining and heating quartz sand, lime, and copper ores to fulfill the needs of artists' palettes. Similarly, both Old and New World artisans combined organic dyes such as indigo with clay minerals to create pigmentlike materials based on the color of the dye. In this way they produced chemical formulations like Maya blue that were unknown in nature. Slow progress in the post-Roman era laid the groundwork for the dramatic discoveries of synthetic pigments and dyes during the nineteenth century. The color industry now is dominated by synthetic materials. Chemists have learned how to make chemical alterations in molecules in order to produce essentially any desired hue and to control the properties of fastness and solubility.

5

GLASS

What Is Glass?

Volcanic furnaces operate at temperatures high enough to melt silicate minerals and create pure molten silica. When the melt flows onto the earth's surface, it cools quickly to form the natural glass known as obsidian (Chapter 1). Such materials are different from crystalline solids, which have an orderly arrangement of atoms and molecules in rows and columns of military precision. Most of this order is lost in freely flowing liquids. When liquids are cooled abruptly, as when volcanic flows reach the surface, they solidify so quickly that the atoms and molecules are not able to arrange themselves into crystalline order. Instead, the disordered arrangement of the liquid is frozen into a rigid material known as glass, which is technically a supercooled liquid.

Silica (SiO_2) is particularly prone to form a glassy phase, but its high melting temperature ($1710°C$) is far above the conditions available in ancient pottery kilns or metal furnaces. Not until humans discovered how to modify silica chemically to lower its melting point could they make glass. Positively charged metal atoms (cations) can insinuate themselves into the silicate system, breaking Si–O bonds and causing the melting point to drop appreciably. The addition of alkali metals as cations (Na^+, K^+) enabled ancient glassmakers to melt silica at the temperatures available to them (under $1000°C$) about four thousand years ago.

Silica (the *former*) came from sand or crushed siliceous stones such as flint, and alkali (the *modifier* or flux) came from mineral or plant sources. Sodium and potassium ions are highly soluble in water, so their presence would lead to instability of the glass unless other, less soluble ions such as calcium (available as lime), magnesium, or aluminum are present as *stabilizers*. The earliest glasses are termed soda-lime, since they were composed largely of 60–70 percent silica, 10–20 percent sodium oxide (Na_2O), and 5–10 percent calcium oxide (CaO, lime) as former, modifier, and stabilizer, respectively. Soda, or soda ash, technically is sodium carbonate (Na_2CO_3), but at high temperatures soda loses a molecule of carbon dioxide (CO_2), to form the oxide (Na_2O), which is present in glass. The raw materials may have been brought together in only two batches, since sand often is calcium-rich from the decay of shells and sources of soda also could contain stabilizers. Oxides of transition metals provided transparent colors when present at levels around 1 percent: blue from cobalt(II) or copper(II), violet to black from manganese(III), pink from manganese(IV), yellow from iron(III), and green from the simultaneous presence of Fe(II) and Fe(III). The Roman numerals indicate the number of positive charges on the metal atom, as many of these metals exist in different charged forms, or *oxidation states,* representing loss of electrons from the neutral atom. In early glass, opacity was achieved when materials were present as undissolved particles that scatter light efficiently: red from copper(I), yellow from lead as its antimonate ($Pb_2Sb_2O_7$), and white from calcium as its antimonate ($Ca_2Sb_2O_7$).

The Precursors of Glass

The genealogy of glass is not clear. It may have been an offspring of the manufacture of metal or glazed pottery. During the refining of metal ores, silicate impurities liquify and separate from the molten metal (Chapter 7). The solidified silicates, or slag, are glassy and would have intrigued an early metalworker with an inquiring mind. Refinement of slag could have eventually led to glass manufacture. Theodore A. Wertime suggested that the earliest known glasses were slaggy lumps from the Mesopotamian sites of Eshunna (twenty-third century B.C.) and Eridu (twenty-first century B.C.).

The alternative scenario has glass evolving from the pottery tradition. From at least the fourth millennium B.C., glaze had been used in Mesopotamia and Egypt on flat ceramics such as tiles (Chapter 3). The chemical composition of early glazes was very similar to that of glass, with alkalis in the glaze serving as the flux to reduce the melting point of silica. Techniques, however, had not been developed to adhere the silica-based glazes to the curved surfaces of clay-based pottery. During the fourth millennium B.C., a procedure was developed to use powdered quartz or silica as the glaze former for rounded objects such as quartz beads. Temperatures available were not hot enough to fuse the quartz in the beads, so they were held together by a very stable surface glaze of modified silica. Because these materials closely resemble glass and preceded it in the areas where true glass first appeared a thousand years later, they are serious candidates as a direct antecedent of glass manufacture. The material is often called faience or Egyptian faience, but its constitution is quite distinct from the tin-glazed earthenware pottery of this name, which is related to majolica and was produced in Europe thousands of years later. The term *glazed quartz* is preferred to faience in the present context.

Chemical analysis showed that the cores of these beads were almost pure quartz. The fine particles were held together in the core with small amounts of soda or lime serving as a modifier. The core invariably was stabilized by a coating of a true soda-lime glaze, usually of a blue color from the presence of copper (hence closely related to the pigment Egyptian blue). Efforts to reproduce the technology by Pamela B. Vandiver and by Michael S. Tite have suggested possible modes of manufacture. For example, the quartz body may have been buried in a mass of the glaze and fired. After firing, the bulk of the glaze was still friable (able to be pulverized), but a layer of vitrified glaze adhered to the surface of the quartz pellet. The recovered objects thus obtained a thick layer of glaze and moreover showed no support marks.

These simple procedures for making siliceous (as opposed to clay) ceramics were used in Mesopotamia, Syria, Egypt, and elsewhere to make beads (Figure 5.1), scarabs, bowls, and small figures. By the Eighteenth Egyptian Dynasty, which began in 1580 B.C., the manufacture of glazed quartz objects peaked, both in terms of the quantity of such materials produced and in terms of their quality, measured for example by the intensity of the blue color. Already in the second millennium B.C., glazed quartz was in use on Crete, and somewhat later in India, mainland Greece, and eventually much of central and western Europe.

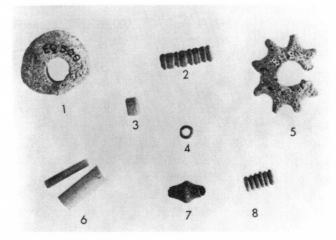

Figure 5.1. Glazed objects of the second millennium B.C. from (1) *Longniddry, Scotland,* (2) *Grimstone Moore, England,* (3) *Kosice, Slovak Republic,* (4) *Branc, Slovak Republic,* (5) *Glenluce, Scotland,* (6) *Egypt,* (7) *North Molton, England, and* (8) *North Molton, England. Bead 2 is 17 mm long.*

Archaeologists originally thought that these materials were manufactured predominantly in Egypt, Crete, and other Mediterranean sites and then were traded up the Rhine or the Danube or across France. Thus glazed quartz beads found in the second millennium in the British Isles were thought to be trade items from Mycenaean Greece. Colin Renfrew first suggested, based on chemical analysis, that local manufacture may have been the rule, even in the British Isles of 1400 B.C. Minoan samples from Crete thus had lower levels of potassium and higher levels of calcium in the quartz cores than did contemporaneous Egyptian samples, and what has been called black faience had higher levels of manganese in the glaze. Glazed quartz beads found in British Middle Bronze Age sites from circa 1400 B.C. also proved to have a distinctive elemental composition. A. Aspinall and his group found that the British beads had almost ten times more tin than did contemporaneous samples from Egypt and elsewhere (Figure 5.2). This unique elemental signature suggested local British manufacture. The tin may have come from the use of recycled bronze (an alloy of copper and tin) by British artisans, whereas fresh copper ores with little tin were used elsewhere. Manufacturing techniques for glazed quartz thus traveled very easily during the second millennium B.C.

Michael S. Tite carried out extensive comparative studies of these and related materials and used a ternary or three-way graph to illustrate their compositions (Figure 5.3). Amounts of the silica former increase from bottom to top along the right edge, of the alkali modifiers from top to bottom along the left edge, and of the calcium oxide (lime) stabilizer from left to right along the bottom edge. Thus pure silica is at the top apex, pure alkali at the left apex,

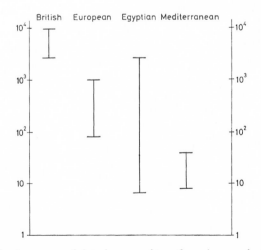

Figure 5.2. The tin content of glazed quartz objects from the second millennium B.C.

and pure lime at the right apex (see Figure 4.5 for further information on ternary plots). Because the core of glazed quartz is almost pure silica, analyses of such materials cluster at the top apex (points labeled + in the figure). True glass contains all three components and is located more centrally in the plot (points labeled by squares containing an x), although slightly to the left side because lime levels are always low. The glaze on the surface of glazed quartz (points labeled by triangles with a vertical line) is seen to have a composition very similar to glass. The plot also contains a few points from Egyptian blue frits (squares containing a short vertical line), which were the raw material for the Egyptian blue pigment (Chapter 4). The pure mineral corresponding to the pigment has the formula $CaO \cdot CuO \cdot 4SiO_2$. With high lime and low alkali, the Egyptian blue points cluster along the upper right edge. The plot shows how Egyptian blue has a composition intermediate between pure silica and glass or glaze.

It is not clear whether metallurgy or the glazed quartz offshoot of pottery was the immediate progenitor of glass. Around 1500 B.C. glass manufacture suddenly emerged as a major industry in Egypt. E. J. Peltenburg sees metallurgy as a more likely ancestor, because both glass and metal manufacture require the raw material to be turned into a molten state. Production of glazed quartz required nothing more than sintering, or heating below the melting

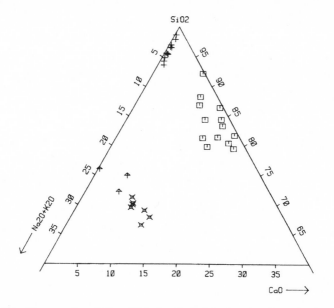

Figure 5.3. Ternary plot of silica (SiO₂), alkali (Na₂O + K₂O), and calcium oxide (CaO) for a variety of early vitreous materials, including the body of glazed quartz (+), Egyptian blue frit (□), glass (×), and the glaze portion of glazed quartz (⩲).

point. Thus a likely genealogy is that pottery led to glazes, including those on quartz, but metallurgy led to glass manufacture. Robert H. Brill, on the other hand, views glazed quartz as the immediate ancestor of glass primarily because the two technologies shared a common composition based on silica. Only further archaeological discoveries and chemical analysis can resolve this uncertainty.

Egyptian Glass

Sporadic finds of glass objects prior to 2000 B.C. have been made in Egypt and even earlier in Mesopotamia. The exact place and time of the discovery of procedures for making true glass and working it into useful or decorative shapes have not been pinpointed. Most scholars think that the discovery was in Mesopotamia but that refinement and exploitation took place in Egypt, particularly after 1500 B.C. during the flowering of the Eighteenth Dynasty, which followed the expulsion of the Hyksos and the founding of the New Kingdom. It was at this time that the first glass containers were produced. A core made of sand or mud was covered with powdered glass frit. The core provided the shape as the glassy raw materials melted during firing. After the piece was fired, the core was dug out to leave a pure glass object. Sometimes residues of the core remained for the modern scientist to detect and analyze. The glass surface could be reheated and decorations applied to produce very attractive colored vessels (Figure 5.4), whose shapes closely resemble those of earlier pottery. This was a landmark achievement in technology, for an entirely new material capable of being shaped and decorated had been produced, thanks to chemical ingenuity. Eventually true casting developed, with molten glass poured into forms. The earliest datable vessels bear the cartouche (circled name hieroglyph) of Tutmosis III from the Eighteenth Dynasty, soon after 1500 B.C. (object on the left in Figure 5.4).

The specific combination of silica, alkali, and lime that these people discovered proved to be particularly successful in the production of glass. Soda-lime glass has come down to the present largely unchanged. Other compositions have been discovered, but soda-lime glass is still superior in terms of ease of manufacture, coloring ability, clarity, and stability to solvents such as water.

Egypt's source of sand has not been determined. It must have had a relatively large amount of iron in it, since even their least colored glass has a light green tint. Classical writers such as Pliny and Josephus mentioned the excellent properties of sand from the mouth of the Belus River on the Syrian coast. Such seashore deposits contain calcium from the decay of seashells, so that the modifier may have been present naturally. More convenient to the Egyptians, of course, were ample supplies of desert sand. Its calcium content may have been enhanced by the decay of limestone bluffs. It is also possible that calcium was introduced from clay pots used initially to melt the sand mix to form a frit. William M. Flinders Petrie described three different types of clay vessels that showed evidence of use in glass manufacture. The ratio of the rare isotope oxygen-18 to the normal isotope oxygen-16 can provide some information about the source of silica. A survey by Brill of numerous types of glass showed one clear outlier: several samples from the eighth-and-seventh-century B.C. Mesopotamian site of Nimrud. The amount of ^{18}O is high enough to suggest that sand was not the source of silica, and the numbers are consistent with either flint or chert. Thus Assyrian glassmakers (modern Iraq) may have used crushed stones for their former rather than sand.

Although the lime stabilizer probably was present as a natural component of the silica, the alkali modifier still had to be added from a distinct source. Egypt has numerous sources of minerals that could have served as the ancient alkali source. One possibility is natron, which is a complex mix of minerals composed primarily of sodium carbonate (soda, Na_2CO_3) and sodium bicarbonate (baking soda, $NaHCO_3$), but also smaller amounts of sodium chloride and sodium sulfate. Copious quantities have been available in Egypt from the

Figure 5.4. Early Egyptian glass objects in the Aegyptische Staatssammlung, Munich. The goblet on the left contains the cartouche of Tuthmosis III (1501–1447 B.C.) (height, 84 mm). The urn on the right is from Tel-el-Amarna, Eighteenth Dynasty (87 mm).

eponymous Wadi Natroun, located between Cairo and Alexandria and used prior to glassmaking for embalming, medical, and detergent purposes. Plant sources of alkali were mentioned in the clay tablets of the Royal Library of Assurbanipal and are thought to correspond to salicornia, a plant that grows in saline environments and contains high levels of sodium. The high levels of magnesium and potassium in early Egyptian glass more closely resemble these plant sources than natron.

Pigments used in transparent glass, such as copper, cobalt, iron, and manganese, for the most part were well known to Egyptians from their use in wall paintings. These materials are fully dissolvable in the glass. Some pigments, however, remain crystalline, and scattering of light from the small crystallites results in a cloudy (opalescent) or opaque glass. This effect was produced in Egypt almost entirely by the use of antimony minerals. Stibnite (antimony sulfide, Sb_2S_3) had been used as a cosmetic for the eyes. When added to glass during its manufacture, stibnite reacted with the lime to form calcium antimonate ($Ca_2Sb_2O_7$), which conveys an opaque white color to the glass. If lead was present, the result was an opaque yellow color from the mineral lead antimonate ($Pb_2Sb_2O_7$).

Experimentation with antimony alone revealed a new property: that it could be used in small quantities to counteract the coloring properties of residual iron. Highly charged antimony (+5 in Sb_2O_5) can pick up negatively charged electrons from iron and be reduced to the +3 oxidation state. When the highly colored +2 iron (Fe(II) as in FeO) is the source of the electrons that go to antimony, the iron is oxidized to the +3 oxidation state (Fe(III) as in Fe_2O_3), which is less highly colored. Thus movement of electrons from iron to antimony results in loss of color. With antimony as a *decolorant,* nearly clear, transparent glass could be produced, although a residual green color from iron (even in modern glass panes) may be observed by viewing the glass along its edge.

In 1961 E. V. Sayre and R. W. Smith reported an intellectual synthesis of glass composition based on their analysis of several hundred samples. They concluded that the five elements—magnesium (Mg), potassium (K), manganese (Mn), antimony (Sb), and lead (Pb)—could distinguish several classes of ancient glass (Figure 5.5). The earliest glass (open diamonds in the figures)— what they called "second-millennium B.C." glass, which includes Egyptian glass—had relatively high levels of magnesium (average 3.6 percent) and potassium (1.1 percent), and very low levels of decolorants such as antimony. They found this high-magnesium glass in use from the earliest times up to about the seventh century B.C. throughout the ancient world, including Egypt, Mesopotamia, and Mycenaean Greece. As mentioned above, the relatively high levels of Mg and K in this glass are consistent with a plant source for the alkali modifier rather than the mineral natron.

About the same time that antimony came into common use as a decolorant, the levels of Mg and K declined (diamonds containing dots in Figure 5.5). The lower levels of Mg and K are consistent with a move away from plant alkalis to the use of sodium-rich minerals such as natron. High-antimony glass was in use from the sixth century B.C. to at least the fourth century A.D. in the eastern Mediterranean world, but not in areas of Roman influence.

By combining glass and metal technology, Egyptian artisans invented the process of enameling. Glass paste of an appropriate color was placed in grooves or depressions on a metal surface. Firing of the object then produced metal with colored glass fused to its surface. Glass would adhere to the metal surface only if it contained enough metal (serving as the colorant) to create metal–metal bonding. Clearly a high level of technology had been attained. Vitreous (glassy) enamel today is used for the surface material of stoves and refrigerators, but to the ancient Egyptians it was an art form that, for example, produced the funeral mask of Tutankhamen, from about 1350 B.C. (Plate 11). Enamel is seen as the decorative blue stripe on the headdress.

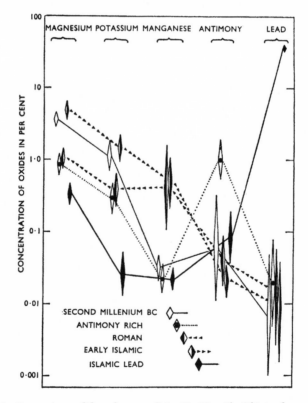

Figure 5.5. Proportions of five elements (Mg, K, Mn, Sb, Pb) in five groupings of ancient glass.

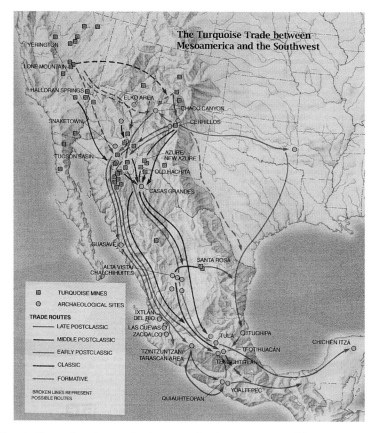

Plate 1. *Turquoise mines in the American Southwest, represented by squares. Proven trade routes are shown with arrows whose color indicates the period of use.*

Plate 2. *Rocks in the Mojave Desert of California, showing the different colors of rock varnish. The dark areas that tend toward black are caused by high levels of manganese, whereas those that are orangish are caused by high levels of iron.*

Plate 3. Red-figured cup (ca. 470 B.C.).

Plate 4. Majolica amphora from Tuscania, Italy (thirteenth–fifteenth centuries), showing a coat of arms.

Plate 5. Porcelain vase with blue underglaze with a design of parrots (Ming Dynasty, first half of the fifteenth century). Reproduced with permission of HarperCollins. MM-1096-65. © John Bartholomew & Son, Edinburgh.

Plate 6. Wool dyed with cochineal is (left) *red with an aluminum mordant,* (middle) *brown with an iron mordant, and* (right) *gray with a copper mordant. Courtesy of Mr. Max Saltzman.*

Plate 7. Egyptian wall painting from the tomb of Pashed. © by 1962 UNESCO.

Plate 8. Fresco from Room 1 at Bonampak, Chiapas, Mexico.

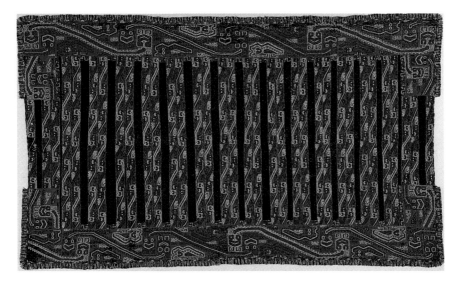

Plate 9. Mantle or cloak from Paracas, Peru, ca. 100 B.C. (98 x 57 inches). Courtesy of the Los Angeles Museum of Art.

Plate 10. Seventeenth-century Chinese sang-de-boeuf vase, Smithsonian National Museum of History and Technology.

Plate 11. The funeral mask of Tutankhamen.

Plate 12. Examples of Roman glass: a striped bowl, a dish with a composite mosaic pattern (15.3 cm), two blown color-band bottles, and a shorter gold-band bottle. Courtesy of the Toledo Museum of Art (Gift of Edward Drummond Libbey).

Plate 13. The Portland vase.

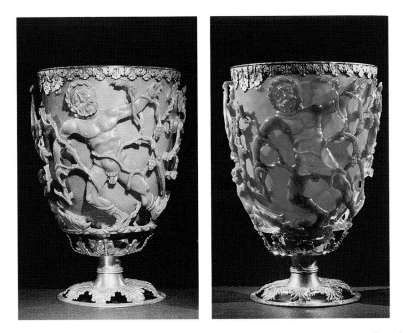

Plate 14. The Lycurgus cup, shown on the left with its green appearance in reflected light and on the right with its red appearance in transmitted light.

Plate 15. Necklace of beads alternating pure tin, amber, and glazed quartz, from about 1400 B.C., found in a bog in Holland.

Plate 16. Peruvian gold cup, Museo Oro del Peru.

Roman Contributions

The first great flourishing of glass production during the middle of the second millennium B.C. in Egypt had already weakened by 1200 B.C. Syria and Mesopotamia provided a revival in the ninth century B.C., and Phoenician traders probably spread the technology to Cyprus, the Aegean, and the Italian peninsula. Egypt saw a resurgence of glassmaking in Hellenistic Alexandria, and both the Syrians and the Alexandrians were instrumental in creating the second great flourishing of glassmaking, which occurred during the Roman era. Classical Greece produced no important contributions to glassmaking, either technical or aesthetic, apparently being satisfied to work in clay. After the Roman conquest of the Hellenistic glass centers of Alexandria in Egypt and Sidon in Syria, Rome provided a new market for glass products. The standard Latin word for glass, *vitrum,* from which many English words derive (vitreous, vitrify, vitrine), was absent from the extant literature before 70 B.C. but was common half a century later. The conquest of Syria by 63 B.C. and of Egypt in 30 B.C. led to the introduction of Syrian and Alexandrian glass and, most likely, the arrival of glassmakers in Rome. By the Augustan age (27 B.C.–A.D. 14) glass had become increasingly common. The Roman geographer Strabo of this period mentions glass manufacture at Rome.

What has become known as Roman glass thus probably was developed in the Syro-Palestianian area by the third century B.C. It had the standard soda-lime content, with magnesium and potassium usually under 1 percent, consistent with a mineral source for the modifier. What set Roman glass apart from its direct predecessors was the use of manganese (Mn) rather than antimony as a decolorant, even though antimony was superior to manganese in this property. Antimony was reintroduced into Italy by the second century A.D. and was used continuously in non-Roman southwestern Asia. By the fifth century A.D. antimony was permanently abandoned in the west after two millennia of use, presumably for reasons of availability or economy. Antimony also was replaced as an opacifying agent by tin oxide, although not until about the fifth century A.D. Roman glass (high Mn, low Mg and K) constitutes the third category found by Sayre and Smith, shown in Figure 5.5 as diamonds with the left half filled in.

The chemical effect of manganese is well understood. When manganese is highly charged (+3 or +4), it creates a violet (Mn(III)) or pink (Mn(IV)) color. In the +2 state (Mn(II)), manganese is less colored. Iron, on the other hand, is very dark in its +2 state (Fe(II)), but is green when Fe(II) and Fe(III) are mixed. As Fe(III), it gives glass a yellow color, or amber when in the presence of sulfides. Reduction of manganese and oxidation of iron, as in Equation 5.1, thus results in a decolorized product, as the highly colored forms of both metals are replaced by the lightly colored forms. Note that manganese goes down in charge (oxidation state) as iron goes up. Anywhere from 0.1–1.6 percent of manganese might have been required for this purpose.

$$Mn^{4+} + 2Fe^{2+} \longrightarrow Mn^{2+} + 2Fe^{3+} \qquad (5.1)$$

The first glass objects produced and used in Rome were formed by the traditional method of casting, often with preformed molds. Sometime around 30 B.C. and probably in Sidon, a glass artisan discovered that glass in its molten state could be inflated or blown into bubbles. The resulting sphere could be made into either flat or elongated shapes, or blown into a reusable mold of almost any desired shape. This discovery not only gave rise to the mass production of high-quality glass products but also enabled new shapes and types of decorations to be developed because of the much thinner material. It is interesting that the ancient soda-lime substance was ideally suited to this new process, even taking excellent impressions when relief was introduced on the inside of the mold. The strong demand in Imperial Rome for both cast- and blown-glass products resulted in the appearance of large manufacturing centers such as Jalame (Jelemie) in Roman Palestine.

The Roman economic expansion was also responsible for the development or extension of many other glassworking techniques. For example, decoration of glass surfaces by cutting had been practiced by the Egyptians, but in Imperial Rome the use of a cutting wheel advanced the art to a point that would not be achieved again until possibly the seventeenth century. The use of thin, flat glass sections for windows also was developed at this time. Excavations at Pompeii, which was destroyed in A.D. 79, have uncovered many samples of window glass, some sections as large as 30 by 40 inches.

Mosaic or millefiori ("thousand flowers") glass had been developed in Alexandria but also was advanced to a high art form in Imperial Rome. Colored glass rods, or canes, were arranged side by side so that their cross section provided a pattern. The canes were fused together and cut into beads or raw material for more complex mosaics, both representational and nonrepresentational. Plate 12 shows mosaics formed by cutting or swirling vessels that were either cast or blown. Figure 5.6 shows the sophistication of design and the shape of the block of canes, whose cross section is nearly identical all the way through. With exquisite conservation of effort, one slice could be combined with another slice turned around to produce a whole that possesses bilateral symmetry, as in the face in Figure 5.6.

Cased glass is another Hellenistic innovation that reached its zenith during the Roman period. Particularly with inflation techniques, glassmakers could prepare an object with one type of glass and then cover or case the object with a second glass of a different composition. In this way objects were prepared with six or more layers. Cameo glass is an example of this technique, whereby a deeply colored glass is dipped while hot into a melt of opaque white glass. The common blue or violet base was obtained primarily by the presence of iron but also cobalt, copper, or manganese. The white opaque layer was achieved by the generous presence of calcium antimonate. Artisans then selectively removed the white material while it was cold to achieve the desired design of white on violet. The method probably was inspired by relief work on banded gemstones such as chalcedony.

The Portland vase is an example of cameo art (Plate 13) and has been said to be the most famous object of ancient glass. It was discovered in 1644 in a sarcophagus three miles from Rome. First kept in the Berberini Library, it was

Figure 5.6. The bar of Roman mosaic glass at the bottom was the source of the upper piece. The lateral grooves illustrate the procedure of assembly from numerous small canes of glass. Courtesy of the Freer Gallery of Art, Smithsonian Institution, Washington, D.C.

purchased in 1770 for the Duchess of Portland and later donated to the British Museum by the Duke of Portland. In 1845 it was smashed by an anarchist and later reconstructed. Part of its fame is due to the effort by Josiah Wedgwood to create a magnificent stoneware copy in 1790 (Figure 3.9). The mythological design may represent the marriage of Thetis, mother of Achilles. She is holding the marriage torch and is situated between Hermes, the messenger of the gods, and Aphrodite, the goddess of love. At one time the vase was considered to be a later copy or forgery. Analysis of the blue body by the British Museum found 65.7 percent silica, 16.2 percent Na_2O, 9.0 percent lime, and only 1.0 percent K_2O, a typical Roman composition. A very similar overall composition, save for the coloring and opacifying agents, for the opaque white layer explains the strong bond between the body and outer layer.

By A.D. 500, this second major flourishing of glassmaking was beginning to wind down as the economic power of the Roman Empire dissipated. Some time during the fourth century, however, a masterpiece of aesthetics and technology was produced in the Lycurgus cup (Plate 14). It was not reported until 1845, and its findspot is unknown. Its composition of 73.5 percent silica, 14.0 percent Na_2O, 6.5 percent lime, and only 0.9 percent K_2O again supports a Roman composition. The vase itself was blown, but the delicate frieze was cut and attached. The figures illustrate the myth of King Lycurgus, who attacked Dionysus and Ambrosia. The goddess Diana transformed Ambrosia into a vine that strangled Lycurgus. This 165-millimeter-tall cup (about 6.5 inches) has the extraordinary property, called *dichroism,* of appearing to be one color (pea green) to reflected light but another color (wine red) to transmitted light, both shown in Plate 14. Robert H. Brill attributed this property to the presence of small particles of gold and silver present in the body. Small amounts of iron, manganese, and antimony could not produce such an effect. D. J. Barber and I. C. Firestone were able to obtain electron micrographs of the coloring particles (Figure 5.7), which have attractive, rounded, polygonal shapes and are made up of 66.2 percent silver, 31.2 percent gold, and 2.6 percent copper. They are 50 to 100 nm in diameter and on the average are 10 μm apart in the glass. (A nanometer, or nm, is a billionth of a meter, and a micrometer, or μm, is a millionth of a meter.)

Both the composition and the particle size were critical in achieving the dichroic effect. Gold alone is known to produce rose to ruby colors in glass and glaze by permitting the transmission only of reds through strong absorption of other colors. Normally this effect, as in the glazed Chinese porcelains called *famille rose,* requires particles of 20 to 60 nm. The green reflected color is associated with the presence of silver and requires particles large enough (usually at least 40 nm) to cause light to scatter back toward the observer. Thus the particles like that shown in Figure 5.7 have just the right amounts of

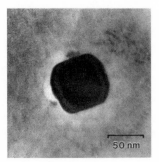

Figure 5.7. Image of a particle of the gold–silver–copper alloy responsible for the dichroic effect of the Lycurgus cup.

gold and silver and are just the right size, fine enough for red transmission and coarse enough for green reflection. The source of this unusual mixture of silver and gold is not known, but Barber and Firestone suggested it was scrap metal from assaying or refining.

Glass in Asia

The last section in this chapter examines post-Roman glassmaking traditions in the West, but it is worthwhile first to examine the earlier and later chemical trails of glassmakers in Asia. The Sayre and Smith plot in Figure 5.5 shows two more categories of glass. The points composed of diamonds with the right half filled are from early Islamic samples. The composition of this glass indicates a return to (or continuation of) the relatively high magnesium (Mg) and potassium (K) levels that reflect plant sources of the alkali. They parallel the first group (open diamonds in Figure 5.5) from the same areas during the second millennium B.C. The Islamic glass, however, also contained high manganese in the Roman fashion as a decolorant. The increase in Mg and K levels occurred sometime in the ninth century A.D., as demonstrated by Sayre's study of Islamic Egyptian and Syrian weights. As Figure 5.8 shows, all samples before about A.D. 845 had the standard Roman composition, but after that date they had the early Islamic composition. These materials, like almost all their predecessors, were composed predominantly of the soda-lime components discovered during the second millennium B.C. by Mesopotamian or Egyptian artisans. The levels of K (1–2 percent) were high compared with Roman glass (under 0.5 percent).

The only alternative to soda as the modifier that had been explored in Europe and southwestern Asia appears to have been lead (Pb). In addition to its coloring and opacifying properties, lead as its oxide can also be used to lower the melting point of silica. It had been used for this purpose in glaze from at least 1700 B.C., when it was mentioned on Babylonian clay tablets. Glass with lead oxide as the flux was manufactured first possibly in Mesopotamia around the seventh century B.C. Lead (up to 35 percent or more of PbO) was found to provide a clear brilliance to glass that was appreciated by Romans and much later was utilized in European crystal and cut glass. Some Islamic glass also used very high levels of lead. The fifth and last class of glass described by Sayre and Smith (the filled diamonds in Figure 5.5), which they call Islamic lead, has an average of 35 percent PbO and, for the first time, low levels of sodium and calcium, as well as of magnesium and potassium. Like the earlier Roman lead glass, however, these examples were rare, the Sayre and Smith group being based on only six analyses. Lead usage also increased in the West, as the standard yellow opacifying agent lead antimonate ($Pb_2Sb_2O_7$) was replaced by cubic lead tin oxide ($PbSnO_3$) by the fifth century A.D.

Lead, however, was a central feature in the earliest Chinese glass. Recent analyses by Brill and by others have found five or more distinct chemical families of Asian glass. The early Chinese glassmakers modified silica with lead and barium rather than with sodium and calcium. Such a dramatic difference in composition almost certainly indicates a separate discovery of glass. From the fifth or sixth centuries B.C. through the end of the Han Dynasty (A.D. 220), East Asian glass was almost always of this unique composition, in one study averaging 47.8 percent SiO_2, 30.4 percent PbO, and 15.3 percent BaO, as well as only 4.03 percent Na_2O and 1.53 percent CaO. If lead was

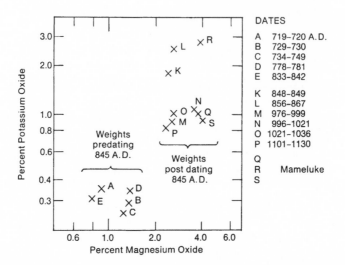

Figure 5.8. Evolution of Mg and K content of Egyptian glass weights between A.D. 700 and 1130.

replacing sodium as the modifier, barium may have been replacing calcium as the stabilizer (both are alkaline earths). This profoundly different lead–barium composition eventually could not compete with soda-lime glass, as it tends to devitrify, producing crystals of barium disilicate, seen as a harsh turbidity. After the fall of the Eastern Han Dynasty in A.D. 220, the lead–barium composition was largely replaced by other formulations. The use of barium would not be seen in Western glass until the late nineteenth century. Some other East Asian glasses, particularly from Japan, were based on lead oxide but lacked barium and hence fortuitously resembled Roman and Islamic compositions.

The influence of the tradition in China of using barium instead of calcium also appeared in the composition of pigments. In Chapter 4 Egyptian blue was described as the first synthetic chemical, discovered before 2000 B.C. As a pure mineral this substance has the formula $CuO \cdot CaO \cdot 4SiO_2$, although the synthetic materials invariably had an excess of silica (see Figure 4.5). Remarkably, during the Han Dynasty, when barium was a significant component of glass, the Chinese discovered a blue pigment with the identical composition as Egyptian blue, except that barium replaced calcium: $CuO \cdot BaO \cdot 4SiO_2$. In addition to Chinese blue, a similar pigment called Chinese purple was known. Analyses by Stephen S. C. Tong revealed that its structure is $CuO \cdot BaO \cdot 2SiO_2$, the lower proportion of silica causing the distinct color variation.

The presence of lead in Chinese and Japanese glass has lent itself to provenance analysis through isotope ratios. Much of this work was carried out by Robert H. Brill, who earlier had identified Western sources from Egypt, Greece, Mesopotamia, Italy, England, and Spain. The earliest Chinese samples have extremely high or low lead isotope ratios, never found in the Western tradition, making the identification of Chinese provenance unambiguous in these cases. His Japanese samples were from Shōsō-in. Although the Japanese lead isotope values overlap with European samples, they correspond to Japanese ore and indicate a source distinct from the two Chinese extremes.

Lead-modified glass was replaced by soda and potash glasses in China after the end of the Han Dynasty. Potash glass is particularly interesting, as it preceded the development of such glass in the West and almost certainly was of Asian origin. In general, Asian potash glass had less than 1 percent of magnesium, in contrast to later Western potash glass. Because of the low levels of magnesium, these Asian glasses probably used purified plant ash or saltpeter as the source of modifiers, according to Brill. Soda-lime glasses of both the natron type (low potassium and magnesium) and the plant ash type (high potassium and magnesium) became increasingly common in East Asia after A.D. 600 and are often indistinguishable from Western glass, although oxygen isotopes may provide some basis for distinction.

In addition to unique differences in the modifiers, Asian glass also revealed innovations in the opacifiers. During the Tang Dynasty (A.D. 618–907) the opacity of jadelike glass was obtained by the use of calcium fluoride (fluorite, CaF_2), which preceded its use in the West for this purpose by almost a thousand years. Fluorite became the principal opacifying agent in the West in the seventeenth century, when it replaced tin oxide, which in turn had replaced the antimony minerals during the fifth century A.D.

From the ninth to the eleventh centuries A.D., Islamic glassmakers developed a procedure for the surface coloring of portions of vessels. A paste of a metal oxide was applied to the transparent glass surface, and the vessel was reheated in a reducing atmosphere. The metal oxides were converted to the free metal, whose small particles migrated into the glass and conveyed color, much as with the Lycurgus cup. The resulting metallic or mirrorlike surface led to the use of the term *lusterware* or luster glass for the product. Silver particles gave rise to yellow or amber colors and copper particles, to orange to red colors. If reduction of copper was arrested at the copper(I) stage (cuprous oxide, Cu_2O), vivid reds could be obtained. Control of concentration, particle size, and extent of reduction required a high degree of expertise.

Luster glass can retain its transparency and hence contrasts with opaque yellow glass, produced by particles of lead antimonate or lead tin oxide, and with opaque red glass, produced by particles of cuprous oxide, often with lead oxide. It also contrasts with enamels (glassy surfaces on metals, as in Plate 11), which were produced in China from at least the Tang Dynasty (A.D. 618–907). The enamel material remained as a distinct phase on the surface after heat treatment, whereas the luster coating actually moved into the glass. The cloisonné technique (Figure 5.9) began in early-fifteenth-century A.D. China and involved the reheating of a glassy paste of metal oxides onto a metal surface. A temperature of 600–700°C normally was required to fuse the glassy material to the metal. The characteristic aspect of cloisonné was the use of thin metal strips, called cloisons, to separate different colors. This technique permitted the construction of well-defined, intricate designs in glass on the surface of metals, usually simple copper alloys.

Julian Henderson and co-workers carried out chemical analysis of Chinese cloisonné enamels and found that the earliest samples, from the first half of the 16th century, had low levels of silica (about 35 percent) and high levels of lead oxide (about 37 percent), as well as appreciable levels of potassium (about 11 percent) and calcium (about 5 percent) oxides. The silica levels increased and the lead levels decreased over the next century. They also analyzed two types of related porcelain enamel (a type of glaze for porcelain) from the eighteenth century. The *famille rose* samples (from 1725) have very similar compositions to the cloisonné enamels, but the *famille verte* samples (about 1700 to 1725) have much higher lead oxide and lower silica than any of the cloisonné

Figure 5.9. Mei-ping vase (Ashmolean Museum, Oxford) with cloisonné enamel.

samples. Examination of opaque cloisonné enamels showed that the opacifying agent was usually calcium fluorite.

Early Non-Roman European Glass

When Sayre and Smith carried out their seminal study of ancient glasses in the early 1960s, they focused on the Mediterranean and Southwest Asian cultures. As a result they did not include all these unique groups from the rest of Asia. In addition, they did not examine non-Mediterranean European samples, other than clearly Roman material. There were in fact indigenous glassmaking activities that both preceded and succeeded the Roman world. Glazed quartz beads from the British Middle Bronze Age, around 1400 B.C., contain enough tin to suggest that they were made locally (Figure 5.2). Similar evidence suggests the local production of glazed quartz in Hungary and Czechoslovakia as well during the second millennium B.C. It is therefore not unexpected that an indigenous production of true glass could have developed in central and western Europe.

If the Europeans had used the same compositions as the Southwest Asians, it would be difficult to identify separate production. Quite early, however, Europeans were using a mixed alkali modifier that had more potassium (about 10 percent) than sodium (about 7 percent), in contrast to the second-millennium B.C. glass of Egypt and Mesopotamia. Both potassium and magnesium in Egypt and Southwest Asia averaged slightly above 1 percent (open diamonds in Figure 5.5), consistent with a plant ash source. As found by both Henderson and Brill, a novel mixed alkali composition was in use during the tenth and ninth centuries B.C. at Frattesina in northern Italy; during the ninth through the seventh centuries at Rathgall in County Wicklow, Ireland; and during the eleventh through ninth centuries at Lake Neuchâtel in Switzerland; as well as at other sites. Figure 5.10 is a plot of potassium versus magnesium, with these samples represented by filled diamonds and triangles and by circles with pluses and crosses inside, concentrated in the shaded area at upper left. Not only do the European samples differ from the Egyptian and Mesopotamian samples in having more potassium than sodium, they are also very low in magnesium, usually less than 1 percent ("LMHK" in the figure signifies low Mg high K.)

Figure 5.10 also shows a very large and widespread group along the bottom of the plot, meaning low potassium (but still significant, usually above 1 percent) and large, variable levels of magnesium (1–6 percent). This composition is very similar to the early Mediterranean and Southwest Asian glasses in Sayre and Smith's "second-millennium B.C." category and is found at sites all over Europe from the thirteenth to the sixth centuries, B.C., including the widespread Hallstatt Culture.

Around the eighth century B.C., the Mediterranean and Southwest Asian glasses moved to a natron source for the alkali and consequently had much lower (generally less than 1 percent) potassium and magnesium ("antimony rich," diamonds with dots, and "Roman," diamonds with the left side filled, in Figure 5.5). Many contemporaneous European glass compositions (such as found at La Tène sites) correspond to this type of material and are found in the lower left corner of Figure 5.10 (expanded in the insert).

From the fourth century B.C. to the first century A.D., Celtic sites all over Europe produced an opaque red glass for use as an enamel on metal. The enameling process was similar to that used by the ancient Egyptians (Plate 11), whereby a glass paste was applied to preshaped sunken areas or cells on the metal surface, and the object was then heated to fuse the enamel to the metal. The bulk glass appears to be of the common soda-lime type, to which were added lead as a flux to facilitate the fusion process and copper to provide the color. These metals were common to Celtic metalworkers. To achieve the bright red color called sealing-wax red, about 25 percent lead and 7 percent copper was used, but the copper had to be the right size, concentration, and oxidation state. In a process probably discovered by the Egyptians and used throughout antiquity, the furnace temperature and oxygen content were controlled to produce cuprous oxide (Cu(I), Cu_2O) crystals of less than 1 millimeter in length. At lower copper and lead concentrations, the Celtic glassmakers could control these conditions to produce elemental copper or cuprous oxide particles one thousandth the size of the cuprous oxide crystals in order to obtain a dull red color, which sometimes was dichroic (red on

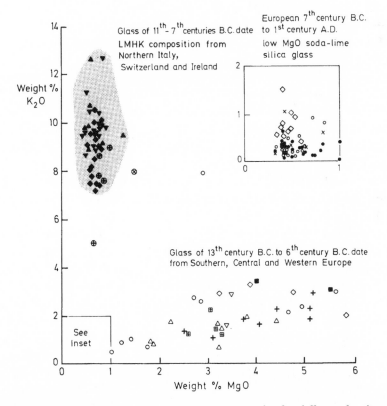

Figure 5.10. Plot of magnesium versus potassium oxides for different families of European glasses. The insert (box) contains points with a composition similar to low antimony and Roman glass in Figure 5.5. The grouping across the bottom is similar to second-millennium B.C. glass in Figure 5.5. The points in the shaded area at the upper left make up a new composition of early European glass.

reflection but green on transmission). Care had to be taken to control conditions not only during manufacture of the glass, but also during the fusion process onto the metal, since copper is easily oxidized to the cupric form (Cu(II)), which typically is green. The Celts used this enamel to decorate torques, brooches, belts, weapons, and even harnesses. Apparently it did not appeal to the Romans or Romanized Celts, as the technique had disappeared by the time Rome conquered the Celtic world during the first century A.D.

Post-Roman Glass from Europe

For the many centuries of Roman domination of southern and central Europe, and for centuries thereafter, European glass had the standard soda-lime composition. Glassmaking centers, known as glass houses, grew up all over the Roman Empire during the first, second, and third centuries A.D., including the Gaulish sites of Boulogne, Amines, Namûr, and Rheims, the German sites of Trier and Cologne, and the British sites of Manchester and Leicester. The founders of these centers may have come from the eastern Mediterranean. For Trier and Cologne it is interesting to imagine a connection between the need for glass bottles and the introduction of wine-making from Italy along the Rhine and Moselle rivers during this same period.

After the decline of the Roman Empire, glass chemistry remained the same for the subsequent Frankish, Merovingian, and Saxon cultures, although ornamentation slowly disappeared. At some time the late Latin word *glesum* came into use, possibly first at Trier from a German source, eventually giving the modern English words *glass* and *glaze*. The eastern Roman Empire, centered at Constantinople, continued the art of decorative glass and probably passed it on to Venice, where in 1279 the guild of glassmakers was formed. Byzantine Jews, possibly from Tyre and Hebron in the ninth century A.D., may have been instrumental in the development of glassmaking at Venice, and they even may have brought sand from the Belus River or from the desert between Cairo and Alexandria for this purpose. Venice in turn stimulated the production of artistic glass throughout Europe.

Medieval glass followed the Roman composition until the end of the tenth century, when a rapid switch occurred, primarily in the northwestern part of the former Roman Empire, from soda to potash glass. (Potash is potassium carbonate, K_2CO_3, which yields the oxide on heating.) Sodium had been the primary modifier everywhere except Eastern Asia since the discovery of glass. Its use required access to minerals such as natron or plants such as salicornia, normally found in the saline environment of coasts. In the absence of reliable trade routes to coastal areas, or because the sodium-rich materials were too costly to import, or maybe just because local plants worked well, glassmakers, around A.D. 1000, turned to inland wood ash sources, such as beech trees, for the alkali modifier. Figure 5.11 is a plot of potassium versus magnesium and shows four major composition types, three already seen. The squares on the left are from Nuzi, a thirteenth-century B.C. Southwest Asian site that has high magnesium and intermediate potassium, characteristic of the earliest glass (the "second-millennium B.C." category of Sayre and Smith). The circles, concentrated in the lower left corner (low potassium and magnesium, characteristic of Roman glass), are from the great glassmaking site of Jalame, in western Galilee. The crosses are the distinctive early European composition, with high potassium and low magnesium, found in this case in Frattesina, Italy (the shaded region of Figure 5.10). Finally, the triangles represent the new composition, from stained glass from St. Maur des Fosses, France (thirteenth century). They average 20 percent potassium with about 5 percent magnesium, the highest potassium levels observed.

The use of potash glass was by no means exclusive after A.D. 1000. In their study of medieval stained glass, J. S. Olin and E. V. Sayre noted that considerable variation in the relative amounts of sodium and potassium in their samples contrasted with observations on more ancient glass. Analysis of late-fifteenth- and early-sixteenth-century Venetian glass by Brill from a shipwreck

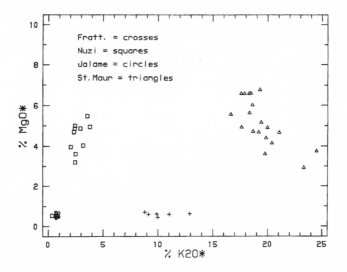

Figure 5.11. Plot of potassium versus magnesium oxides for glass from Frattesina in northern Italy (early European group of Figure 5.10), Nuzi in Western Asia (second-millennium B.C. in Figures 5.5 and 5.10), Jalame in Palestine (Roman in Figures 5.5 and 5.10), and St. Maur in France (a new composition of Medieval potash glass). Note that the axes in this figure are reversed from those in Figure 5.10. Courtesy of Robert H. Brill, The Corning Museum of Glass.

found only soda glass. The Venetians were continuing the ancient Mediterranean tradition. The use of potash glass by medieval glassmakers has had one unfortunate result: Because it is less stable than soda and mixed alkali glasses, it is more prone to weathering in the atmosphere and to disintegration during burial.

Analysis of the relative amounts of potassium and magnesium has been taken as a measure of the plant source. Figure 5.12 shows a series of lines that represent the K/Mg ratio for specific plants. Norman Tenant and his colleagues analyzed thirteenth-century glass from Scottish cathedrals and placed them on this plot. Most points fell between the lines for fern and beech, so they concluded that the alkali came from a mixture of fern and beech ash. Their choice of beech may have been influenced by the existence of an early-twelfth-century treatise by Theophilus, a Benedictine monk and practicing metal- and glassworker who recommended the use of beech ash, and by Eraclius, of the same or earlier time, who actually described the fern-beech combination. At least the modern data are in accord with these medieval recipes.

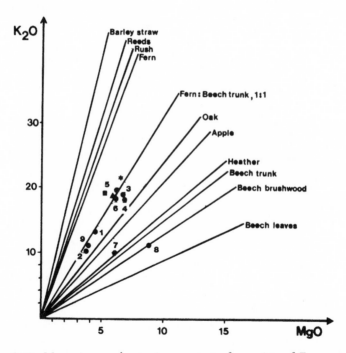

Figure 5.12. Magnesium and potassium content of a variety of European plants (lines) and of Scottish glass from Elgin Cathedral (circles), Holyrood Abbey (triangles), St. Andrews Cathedral (stars), and Melrose Abbey (squares). Reprinted with permission. © 1984 by the American Chemical Society.

There were very few medieval chemical innovations in glassmaking. The standard minerals were used to produce the colors for stained glass: cobalt for blue, copper and iron for greens and reds, manganese for purple, tin for white. Enameling techniques followed ancient traditions. During the eleventh and twelfth centuries in Limoges, France, delicate enamels were produced by fusing colored glass to areas that had been scraped out of pure copper, the *champlevé* ("raised field") method, similar to the ancient Egyptian method (Plate 11). The Venetians later developed techniques to produce extremely thin glass vessels, called *cristallo,* but not by chemical changes. In 1674 George Ravenscroft in England discovered or rediscovered the brilliance conveyed by high levels of lead in glass, and his glass house, which used about 15 percent lead, created modern lead crystal, which supplanted the Venetian styles. The term *crystal* is somewhat of a misnomer, since glass by definition is noncrystalline. Some minerals, including quartz, occur as brilliantly clear and colorless crystals. The resemblance of glass to these materials provided the name.

This chapter has described only a few glass compositions, based primarily on the properties of the modifiers, stabilizers, and decolorants. Despite the passage of almost four millennia, the fusion of silica at a reasonable temperature by the addition of alkali has remained the technological cornerstone of glassmaking. Although plastics have replaced glass in specific contexts, and some modern discoveries, such as the use of boron, have occurred, soda glass remains the standard, and it would be immediately recognized by a time-traveling Egyptian or Roman glassmaker.

6

ORGANICS

Organic Artifacts

Fire was probably the first chemical reaction regularly carried out by humans. Observation of natural fires and experimentation with fuels led to this first chemical manipulation of the environment. Controlled fire provided warmth, protection from predators, and a means to prolong the use of certain types of food. Controversial evidence for fire has been found in the South African Swartkrans cave dating back over a million years. C. K. Brain and A. Sillen found chemical traces of char on bones, along with skeletal remains of two species antecedent to humans, *Australopithecus robustus* and *Homo habilis.* The frequency of the charred remains indicated that fire was a common event. Stronger evidence confirms that fire was used by *Homo erectus,* the predecessor of modern humans *(Homo sapiens),* at least 200,000 years ago. Regularly transforming matter, these early hominids had become chemists.

Fire involves the conversion of carbon and hydrogen in a fuel to carbon dioxide (CO_2) and water (H_2O). Energy is given off in the form of light and heat, as the more stable molecules are formed. Natural fuels like wood contain large amounts of the element carbon, the source of the product carbon dioxide, and hence belong to the class of compounds that comprise *organic chemistry.* Organic compounds originally included only molecules derived from biological sources, but chemists have created millions of synthetic organic compounds since the founding of the field in the 1820s, so that the word *organic* has lost its original connotation.

Organic compounds make up the food we eat, the clothes we wear, and the wood we burn and build with, as well as numerous other materials, from perfume to paper. Unfortunately fungi and bacteria are able to feed on carbon-based materials, which tend to be much shorter lived under archaeological conditions than inorganic materials such as stone, pottery, glass, or metals. When they do not disappear entirely from the archaeological site, their remains usually are quite fragile. Moreover, the limited number of elements present (usually just carbon, hydrogen, oxygen, and nitrogen) means that the workhorse of previous chapters, elemental analysis, loses some of its power.

Scientists have taken two approaches for the analysis of archaeological organics. On the one hand the complex mix of organic compounds in an artifact may be separated by chromatography and each component identified by mass spectrometry (MS) or other methods. Important molecules sometimes are able to serve as markers for the function or source of the original material. Ultraviolet (UV) and infrared (IR) spectroscopies were used to identify the organic molecules in dyes in Chapter 4. In the second approach, the entire organic material may be examined in bulk directly by IR or nuclear magnetic resonance (NMR) spectroscopies, which can give information about the atoms and bonds present in the molecules. The spectra of archaeological mixtures often can serve as distinctive fingerprints for comparison, in the same way that the pattern of inorganic elements is used as a fingerprint for stone and pottery. This chapter describes the analysis of a number of organic artifacts at the molecular level using these two approaches, in order to gain information about ancient cultures.

Food

Food is of central interest to the archaeologist, since it defines much of an ancient people's culture or lifestyle. Fire was used to transform many types of food chemically for tens of thousands of years prior to the appearance of farming. Cooking made food safer by destroying bacteria, and it possibly made it better tasting and easier to digest, by breaking up large molecules such as proteins and carbohydrates. The cultural conversion from hunting and gathering to farming arose independently at several times and places, but first occurred about ten thousand years ago in Southwest Asia. There is evidence for flint sickles and grinding stones before 8000 B.C. in the semiarid mountains close to the valleys of Mesopotamia. After a thousand years of collecting wild grain, humans moved to the cultivation of grain and domestication of animals. For example, at the site of Jarmo in Iraq, Robert J. Braidwood found seeds of wheat and barley and bones of goats that dated to 6750 B.C. Rice was domesticated in Southeast Asia by 5000 B.C., and possibly earlier. Also by 5000 B.C., maize, squash, and other plants were domesticated in Mexico, and beans and pepper in Peru. Chemistry entered when people began to develop prepared foods. For example, yeast added to a paste of grain gave off carbon dioxide gas, which was trapped to give the lighter material we call bread.

Microscopic examination of archaeological soil can give evidence for individual plant species by identification of seeds, pollen, phytoliths (silica bodies from plants), or even pieces of soft tissue; and for animal species by identification of bone, teeth, or hide. The presence of such materials does not always imply that they were part of the diet. Chapter 8 describes how chemical examination of human remains provides more direct evidence for specific components of the diet.

The process of cooking can leave behind a molecular imprint of the menu. Pottery used in cooking frequently retains dark stains that are residues of food, still present after hundreds or thousands of years. Although the food itself is not intact, different types of molecules can indicate what class of food was cooked. The procedure involves removing the residue from the pottery by solvent extraction or by heating. The resulting mixture of dozens or hundreds of molecules is separated into components by chromatography, and the components often may be identified by spectroscopic tests. In gas chromatography (GC), for example, the mixture travels as a gas through a long column containing a solid that slows some of the gas molecules more than others, by chemical or physical interactions between the gas and the solid. The gases exit from the column sequentially rather than all at once and pass over a detector that registers a peak for each type of molecule. Figure 6.1 shows such an experiment by T. Oudemans and S. Boon on residues from the inside of a pot from a Late Iron Age or Early Roman site called Uitgeest-Groot Dorregeest in the province of Noord-Holland. They found molecules like pyrrole (peak 32) and toluene (88) that are markers for proteins (specifically, for the amino acids proline and phenylalanine), others (11 and 25) like furan that are markers for sugars (another name for carbohydrates), and organic acids (199 and 213) that are markers for fats, waxes, or oils. This variety of compound classes indicates a very nonspecific use for the pot in food preparation.

More specific conclusions came from examination of pottery fragments from a coastal site on the Vredenburg Peninsula at Kasteelberg, South Africa. Fatty acids, which are components of fat, were identified by comparison with authentic materials. The chemists found that the ratio of specific fatty acids (palmitic acid to stearic acid) indicated animal rather than plant origin. Further examination of the ratio of oleic acid and vaccenic acid was consistent with marine animals such as the Cape fur seal, whose bones also were found at the site. They concluded that the pots were used for boiling the meat of a marine animal, possibly seal.

The group of Richard Evershed has actively investigated numerous food residues. They have used the steroidal molecules cholesterol and solesterol as the respective markers for animal and plant foods. From unglazed Late Saxon/ Early Medieval vessels excavated at the West Cotton site in Northamptonshire,

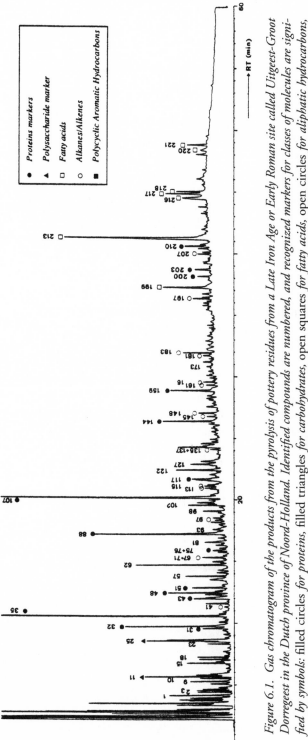

Figure 6.1. Gas chromatogram of the products from the pyrolysis of pottery residues from a Late Iron Age or Early Roman site called Uitgeest-Groot Dorregeest in the Dutch province of Noord-Holland. Identified compounds are numbered, and recognized markers for classes of molecules are signified by symbols: filled circles for proteins, filled triangles for carbohydrates, open squares for fatty acids, open circles for aliphatic hydrocarbons, and filled squares for aromatic hydrocarbons. Reprinted with the kind permission of Elsevier Science-NL, Sara Burgerhartstraat 25, 1055 KV Amsterdam, The Netherlands.

England, they identified nonacosane, nonacosan-15-one, and nonacosan-15-ol, which they associated with leafy vegetables such as cabbage or turnip greens. They also studied the residues on different parts of a vessel (the rim, the body, the base) to distinguish pots used for storage from those used for cooking. For example, they concluded that high concentrations of fatty molecules around the rim are consistent with the boiling of vegetable and meat products, during which fatty materials concentrated on the surface and seeped into the ceramic.

Rather than examining individual molecules, the group of Michael J. DeNiro looked at isotopic ratios of carbon and nitrogen of the mixtures. The ratio of ^{15}N to ^{14}N from a food residue can distinguish between legumes (such as beans) and nonlegumes. The ratio of ^{13}C to ^{12}C can distinguish between plants such as maize, which are formed via a photosynthetic pathway with an intermediate containing four carbons (C_4 plants), and those formed via an intermediate containing three carbons (C_3 plants). Since all legumes are C_3 plants, this method can divide plant residues into three groups: legumes, C_4 plants, and nonlegume C_3 plants. In most cases, the only relevant C_4 plant is maize, also called corn in the Americas.

Figure 6.2 is a characteristic graph, with the horizontal axis the proportion of ^{13}C in parts per thousand and the vertical axis the proportion of ^{15}N in parts per thousand. The three boxes from previous work indicate where to expect the three types of food to fall on the plot. DeNiro, with Christine A. Hastorf, examined residues from the Upper Mantaro Valley in the Peruvian

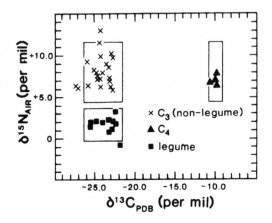

Figure 6.2. Plot of the ratio of ^{13}C to ^{12}C (horizontal axis) versus the ratio of ^{15}N to ^{14}N (vertical axis) in parts per thousand, for plant residues in ceramics from the Upper Mantaro Valley of Peru. The boxes are the expected positions, based on previous work, for legumes, C_4 plants, and nonlegume C_3 plants. The various points are from the analysis of the residues.

Andes (200 B.C. to A.D. 1470). Analysis of residues indicated that the jars were used to cook nonlegumous C_3 plants or C_4 plants, but not to cook legumes.

A third procedure was used by G. C. Hillman to identify the source of charred seeds by spectroscopy when microscopy failed. Seeds from Tell Abu Hureyra and Tell Mureybit in Syria were soaked in solvent. The extracts were analyzed by IR spectroscopy and compared with results from modern seeds. Figure 6.3 shows the close similarity between the spectra of a charred grain from Tell Abu Hureyra and that of *Secale montanum,* a wild rye.

Beer, Wine, and Chocolate

Chemistry was essential in the development of any alternative to water as a drink staple. Farmers require considerable amounts of water to replace losses during the workday. Healthful sources of water have not always been available, and disposal of human wastes often was carried out close to or even in the water supply. Consequently, methods for preserving or detoxifying water would have been very useful. One of these was the addition of ethanol (ethyl alcohol) to water through natural fermentation. Many microorganisms cannot exist in the presence of ethanol. Any substance containing sugars can ferment naturally, because wild yeasts may be present and provide the enzyme required to convert sugar to ethanol.

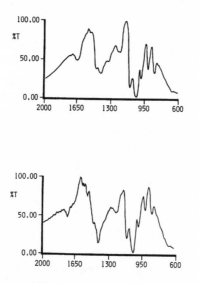

Figure 6.3. Infrared spectra of propanol extracts of charred seeds (bottom) *from Tell Abu Hureyra in Syria and of modern grains* (top) *of the species* Secale montanum, *found in the Munzer Mountains.*

Possibly before the domestication of grain, humans discovered that yeast could be isolated and added to various materials to hasten the process of fermentation. Charles B. Heiser even suggested the possibility that yeast was cultivated before grain, if early brewers retained yeast-containing residues made from wild ingredients as starters for new batches of beer. In the 1950s Robert J. Braidwood, who had carried out the excavations at Jarmo, suggested that bread-making and the domestication of grains were interconnected developments. Jonathan D. Sauer, a botanist at the University of Wisconsin, countered with the suggestion that cereals may have been domesticated initially for making beer rather than bread. A symposium, titled "Did Man Once Live by Beer Alone?," then was convened, although the results were inconclusive as to whether beer rather than bread provided a greater push to the domestication of grain. More recently, Solomon H. Katz and Fritz Maytag offered more evidence on the side of beer. Using a Sumerian poem, "Hymn to Ninkasi," which contains a recipe for beer, they tried to recreate the ancient beverage. Sumerian bread, called *bappir,* was one of the ingredients, suggesting that bread was a means of storing the raw materials (starch) for beer.

In general, early prepared foods offered lower toxicity and improved nutritional values. During the fermentation process to make beer, the levels of stomach-irritating tannins are decreased and the levels of the B family of vitamins and of essential amino acids are increased. Beer provides protein that can be a food staple as well as an intoxicant. Sanborn C. Brown estimated that the farmers in the hills of Tibet today obtain between a third and a half of their daily caloric input from beer.

Beer comes primarily from barley (*Hordeum vulgare* L.), certainly giving one pause to consider why, of all the possible wild grains, barley was one of the first to be domesticated, if not the first. Barley has an unusual, if not unique, property: when the grain germinates, enzymes are produced that break down starch into its component sugar monomers for seedling growth. Starch is a polymer consisting of many sugar monomers linked together, but yeast can operate only on the unlinked sugar monomers. The process of *malting* allows the grain to germinate, release the enzyme, and begin to break the starch polymers down to the sugar monomers. The ground malt is then brewed by being boiled in water *(mashing)* to complete the release of sugar monomers and to form wort. This liquid is sterilized by boiling. Finally, fermentation is begun by the addition of yeast to the cooled wort. Different strains of yeast produce ale and lager beer. Although all the fermented products of barley historically are beer, today we reserve that term for the beverage that is brewed in the presence of hops, which makes a more bitter taste, but also improves the storage properties. The process was introduced by the Dutch in the fifteenth century.

The level of alcohol present is determined by the amount of time the malt is heated and also whether extra sugar is added. The chemical equation for the conversion of sugar to alcohol (Equation 6.1) was first determined by Joseph-Louis Gay-Lussac in 1810:

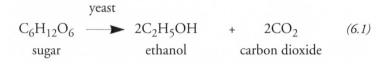

$$C_6H_{12}O_6 \xrightarrow{\text{yeast}} 2C_2H_5OH \quad + \quad 2CO_2 \qquad (6.1)$$

sugar ethanol carbon dioxide

Fermentation is actually a much more complex biochemical process, taking many steps and requiring up to a dozen enzymes and cofactors such as metals. For beer-making, ethanol is the primary product and carbon dioxide is a by-product. The gas creates the head or froth on beer and originally was utilized to assist in preservation. A cask sealed and retained under the pressure of carbon dioxide keeps air and microbes out of the empty portion of the keg (the ullage). In the formation of bread, carbon dioxide is the primary product and ethanol is lost during baking. The gas is caught in the glutinous material, causing it to rise or leaven and produce a lighter and tastier food.

Like farming, the consumption of beer first appeared in or near Mesopotamia, possibly as early as 6000 B.C. The code of Hammurabi in the eighteenth century B.C. imposed stiff penalties on vendors who sold weak beer, overcharged their customers, permitted criminals to frequent their establishments, or allowed political conspiracies to be created on their premises. The "Hymn to Ninkasi" was written about this same time in praise of the Sumerian goddess of brewing and has been found on tablets at Nippur, Sippar, and Larsa. A Sumerian proverb stated, "He who knows not beer knows not what is good."

Babylonian texts document some seventy varieties of beer, and Hittite texts suggest that beer was second in importance only to bread. Pottery from ancient Phrygia from the eighth and seventh centuries B.C. in Anatolia (modern Turkey) indicates that brewing left barley husks and other solid matter in the beer. To avoid this material, beer was drunk either through a straw from an open vessel (Figure 6.4, top) or from a jug with a spout attached to the main body through a sieve (Figure 6.4, bottom).

Both bread and beer were staples of the ancient Egyptians, who left numerous samples in tombs for consumption by the deceased in the afterlife. Many of these samples survived in the dry climate and may be analyzed today. Delwen Samuel carried out a study of nearly seventy remnants of bread loaves and of the contents of some two-hundred pottery vessels thought to contain beer, dating to the period from 2000 to 1200 B.C. Both beer and bread could be identified by their biological and chemical residues. She found that the microstructures of the residues were remarkably similar to those of modern cereal foods. The bread was a baked product from a moist dough, as judged by the expansion of the starch particles that occurred on wetting. Some bread residues contained hollow starch granules, as should occur by the action of enzymes from yeast during rising. The starch particles from the beer residues had a range of microstructures that suggested that well-heated malt (deliberately germinated grain) was mixed with unheated malt. Bread could not have been the source of such grain. Because considerable starch remained in the beer residues, the Egyptian brewing process was very inefficient by modern standards.

Fermented beverages may be made from a variety of other sugar-containing materials, including cider from apples, sake from rice, mead from honey, and,

Figure 6.4. Top, *neck-rim fragment of a Phrygian vessel showing use of a long straw for sipping liquids from a container on the drinker's left.* Bottom, *spouted sieve jug from Tumulus W (740–730 B.C.) in the Phrygian capital Gordion, the city of Midas.*

of course, wine from grapes. Wine became the drink of choice for the Greeks, Romans, and Celts, in part because many areas of Europe were well suited for the growing of the grape, *Vitis vinifera*. As with beer, wine production first appeared in Southwest Asia, probably in the mountainous region between the Black Sea and the Caspian Sea (Figure 6.5), before 5000 B.C. Even as early as 10,000 B.C., the consumption of fortuitously fermented grape juice may have occurred, but it is better to associate its regular use with the cultivation of the vine (viticulture). From its source, wine-making moved to the cultures of the Sumerians, Assyrians, Babylonians, and Hittites. Ancient Egypt produced and imported wine, primarily as a drink of the aristocratic classes. Tomb paintings have provided an extensive record of their viticulture. From Southwest Asia, wine production moved into Greece, Italy, and Gaul, where it became a beverage used by all levels of society. Its trade had enormous economic significance during the Roman era. The wine amphora became one of the most common ceramics of the time. André Tchernia estimated that 40 million amphorae of wine were imported into Celtic territories during a single century at the end of the Iron Age, corresponding to 2.65 million gallons of wine per year, just for the Celts.

About 20 percent of the wine grape is a mixture of two types of sugar, glucose and fructose. Unlike the case with barley, the sugars in grapes already are present as monomers rather than as the polymer starch, so no malting is required. Grape sugars ferment in the presence of yeast by the equation (on p. 135) elucidated by Gay-Lussac. White wine is produced from the clear juice and red wine from juice still containing the grape skins. Originally, fermentation depended on yeasts brought in on the grapes or present in the winery. Since the time of Louis Pasteur, a pure culture of yeast has been added to sterile grape juice, so that the flavor and odor may be controlled more closely.

Natural organic acids such as malic and tartaric acids compose about 1 percent of the grape. Such molecules provide markers in the analysis of residues in amphorae. A dark stain in a pre–Bronze Age jar from Godin Tepe in Iran from about 3000 B.C. was found by Patrick McGovern and Rudolph Michel to contain tartaric acid derivatives consistent with wine as the original contents of the jar. The earliest evidence for wine consumption, also discovered by McGovern, was the presence of tartaric acid in a neolithic jar from the site of Hajji Firuz Tepe, in Iran's northern Zagros Mountains. This site dates to 5400–5000 B.C.

Chemistry has had some impact on the analysis of unfermented drinks as well, playing a small role in the decipherment of Maya hieroglyphics. Enormous strides were made in understanding this picture writing, after it was realized that the language was represented syllabically by glyphs. Glyphs found on elite Maya ceramics often contained information about the original owner and the intended contents of vessels. David Stuart recognized that the glyph shown in

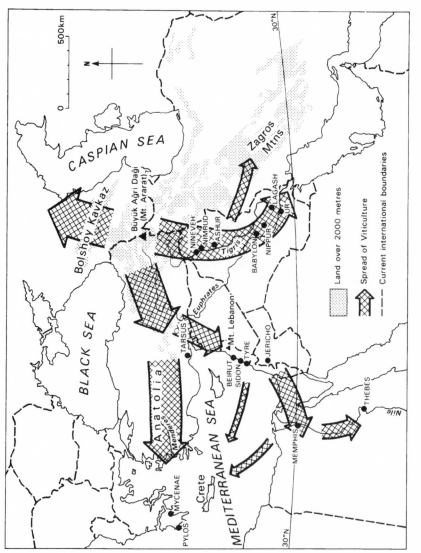

Figure 6.5. Map showing the origins of viticulture before 5000 B.C. in the region between the Black and Caspian seas, and its spread throughout Southwest Asia and the eastern Mediterranean.

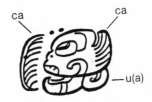

Figure 6.6. The glyph cacao or chocolate: ca-ca-u(a). Drawing by D. Stuart.

Figure 6.6 stands for "chocolate," when he put the three sounds into the right sequence. The comb shape at the left was known to stand for the syllable *ca*, and the fish figure comprising the upper right two-thirds also stood for *ca*. The piece at the bottom right was a final *u*, making the whole the eminently recognizable word *ca-ca-u*. As if in confirmation, in 1984, a drinking vessel was found at Río Azul in the Guatemalan Petén region with the owner's name, the signs for "his drinking vessel," and the glyph for "cacao." The vessel contained enough residue to be analyzed chemically by the Hershey Foods Corporation. It proved to be chocolate: chemistry supporting linguistics. The cacao plant *(Theobroma cacao)* was domesticated in Mexico long before the Spanish conquest. The seeds are fermented, dried, roasted, and ground. The whole bean produces chocolate, containing about 40 percent fatty oils, 15 percent starch, and 15 percent protein. Removal of most of the oils as cocoa butter results in cocoa.

Fibers and Textiles

The construction of fabrics, along with the control of fire and the domestication of plants and animals, must be considered critical to the rise of civilization. The plant or animal sources of the raw material for fabrics had to be discovered and possibly domesticated. From the raw material, relatively short fibers were produced by both physical and chemical processing. From these fibers long bundles, or yarn, were formed by spinning, twisting, or (in the case of silk) throwing. Fabric construction was achieved by interlacing yarn at right angles (weaving), looping and interlooping (knitting), or knotting (lace). Unfortunately, all types of fabric are subject to rapid decomposition through the combined effects of bacteria, fungi, and insects. The earliest evidence for the use of twisted fibers, however, is extremely ancient. Female representations in stone, called Venus figures, sometimes are portrayed as wearing a type of string skirt (Figure 6.7). The illustrated example was found in Lespugue, France, and dates to about 20,000 B.C., from the Gravettian Culture. The carving clearly shows the fraying out of the twisted fibers at their ends. Actual examples of twisted fibers dating to about 15,000 B.C. have been found in the

Lascaux caves in France. Thus the concept of constructed fabric is very ancient, but domestication of the source of the raw materials was necessary before the process could become widespread.

Wool may have been the first fiber used for fabric construction, although there is little primary evidence. It was first produced in Southwest Asia, and possible examples have been found at Çatal Hüyük from around 6000 B.C. The earliest actual wool artifacts date from the fourth millennium B.C. from predynastic Egypt. Sheep had been domesticated as early as 8000 B.C. and provided the first wool, and goats such as the cashmere also were used. In South America the camelids (llama, alpaca, vicuña, guanaco) were used extensively as a source of fiber. Wool fiber is composed of the protein keratin. The fleece is removed by clipping or pulling, scoured (the only chemical step), picked to eliminate extraneous matter, carded to align the fibers, and finally spun into yarn. Originally a product of Mesopotamia, wool became the fabric of choice for Classical Greece and Imperial Rome.

The major competitor for the earliest fabric is linen, which is produced from flax (*Linum usitatissimum,* a domestic annual). Actual pieces of linen textiles have been found at Naḥal Ḥemar in Israel from the seventh millennium B.C.,

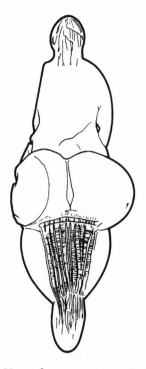

Figure 6.7. Stone Paleolithic Venus figure wearing a skirt of twisted string, ca. 20,000 B.C., Lespugue, France. Collection Musée de l'Homme, Paris.

and at Çatal Hüyük in Anatolia from about 6000 B.C. Linen was widely used in Mesopotamia and Egypt, in the latter case more than wool. Flax is an example of a bast fiber—that is, one that comes from the inner bark of the stem of certain plants. Flax is harvested, stored to dry, retted to remove gummy material, pounded to remove the fiber from the stem, combed or hackeled to organize the fibers, and spun. Retting (from an old form of the verb *rot*) is carried out by exposing the fiber to moisture or water (in the field with dew, in ponds or streams, or on rooftops).

Cotton was first cultivated in India and Egypt, at least by 3500 B.C. From the great city of Mohenjo-Daro in Pakistan numerous examples of cotton fabrics have been dated to about 3000 B.C. Cotton was found in Assyria by 700 B.C., and in Greece by A.D. 200. Cotton was discovered independently in Peru up to five thousand years ago. The site of Huaca Prieta, dating from 2500 B.C., has yielded more than nine thousand textile fragments from desiccated mummy wrappings and funeral offerings. Cotton comes from several members of the genus *Gossypium*. The cotton is removed from the plant, separated as lint from the seeds, carded to align the fibers, and spun.

The discovery of silk has been dated to about 2600 B.C. in China, although physical evidence is lacking until much later. The existence of long-distance trade, eventually associated with the Silk Road, is supported by the introduction of silk to Europe as early as the time of Alexander the Great. China had a monopoly on its production until about A.D. 200, when the Japanese were able to work out the secrets of silk production, or sericulture. The earliest silkworm in China was *Bombyx mori*. The cocoon is harvested before the moth emerges. The silk filament is unwound, or reeled, usually several strands at a time, to attain a workable diameter. As one filament nears its end, another is attached to it. Reeled silk is transformed into yarn by the process called throwing, analogous to spinning.

Other textiles include bast fibers such as hemp (from *Cannabis sativa*), jute, and ramie, and leaf fibers such as sisal and esparto. Figure 6.8 shows the geographical ranges of various prehistoric fibers.

Preservation of fibers requires unusual conditions to prohibit their destruction by insects and microorganisms. Natural refrigeration has been effective, as in the case of woolen materials in Scythian tombs from 500–300 B.C. in the Eurasian steppes, or with the Ice Man from about 3000 B.C., discovered on the border of Austria and Italy in 1993. The absence of oxygen in lakes in Switzerland provided dramatic discoveries of flax products from 3000 B.C. The dryness of Egyptian and Chinese tombs, the Judean desert (Naḥal Hemar), and the Anatolian plateau (Çatal Hüyük), and the peculiar chemical environment of peat bogs in Scandinavia also have provided early or well-preserved examples.

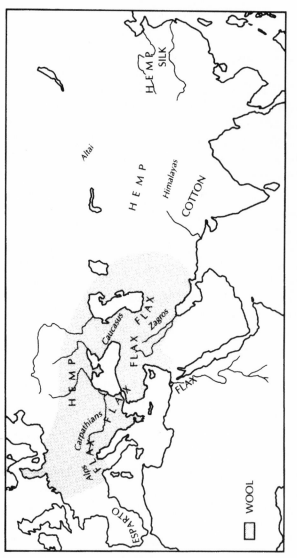

Figure 6.8. Map of the principal fibers used during the third millennium B.C. The area for wool is lightly shaded and extends from the North Sea to India. Reproduced with permission of Princeton University Press. © 1991 by Princeton University Press.

Figure 6.9. Top, *textile impressions on clay from Jarmo in northeastern Iraq, ca. 7000 B.C., with plain weave* (left) *and basket weave* (right). Bottom, *diagram of plain weave or tabby* (left) *and basket weave* (right). *Reproduced with permission of Princeton University Press.* © *1991 by Princeton University Press.*

The earliest proof of weaving comes not from the textiles themselves but from the impressions they left on clay (Figure 6.9). Examples both of plain weaving (tabby) and of basket weave were discovered from circa 7000 B.C. at Jarmo in Iraq. Similarly, the earliest examples of silk are not of the fabric but of a fossilized form. Kathryn A. Jakes and Lucy R. Sibley explained patterns found on the surface of a Chinese bronze halberd (a long-handled battle-ax or pike) as resulting from the molecular replacement of the organic silk with a structure composed of copper oxides from corrosion of the halberd. The object dated to the Shang Dynasty, approximately 1300 B.C. They called this silk-turned-into-copper a pseudomorph.

Fibers may be identified according to their biological source by chemical and microscopic tests, which, unfortunately, are usually destructive. Plant fibers as a group (flax, cotton, and the like) are made up of the polymer cellulose, which can be dissolved, for example, in 70 percent sulfuric acid. Animal fibers (wool, silk) are composed of proteins, which may be dissolved in 5 percent sodium hypochlorite solution (NaOCl). Scanning electron microscopy (SEM) normally is the technique of choice for distinguishing most fibers, by comparison with reference standards for the cross section and the longitude of the fiber. The material may have to be cleaned chemically (organic solvents for

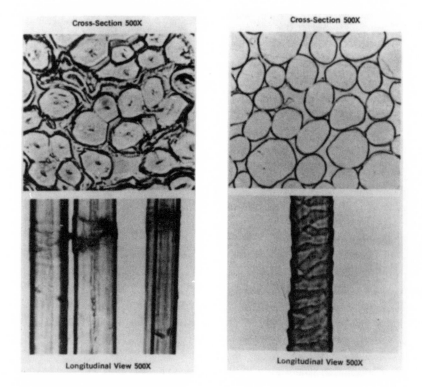

Figure 6.10. Left, *scanning electron microscopic view of flax in cross section and longitude.* Right, *scanning electron microscopic view of wool in cross section and longitude.*

the removal of oils and waxes; hydrosulfite caustic solution for the removal of dyes). A thin section is removed and set into a matrix for the SEM experiment. Figure 6.10 illustrates the differences between wool and flax in the two views.

A team from the University of Illinois at Urbana-Champaign carried out analyses of this sort on an Egyptian mummy in the collection of their World Heritage Museum, dating to the Roman period (first to second century A.D.). Samples from the wrappings dissolved in 70 percent sulfuric acid and hence were a cellulose fiber. The SEM examination of the middle and inner fabrics revealed the morphology typical of flax. The outer fabric, however, lacked the longitudinal nodes of flax that are apparent in Figure 6.10. The team concluded that the material is ramie, a bast fiber from nettle plants known to be highly resistant to insects and microorganisms and hence attractive to the embalmer as a protective covering for the mummy.

One of the most remarkable recent discoveries of textile material occurred in 1986 at an excavation in the northern Arabian site of Dhuweila (Figure 6.11),

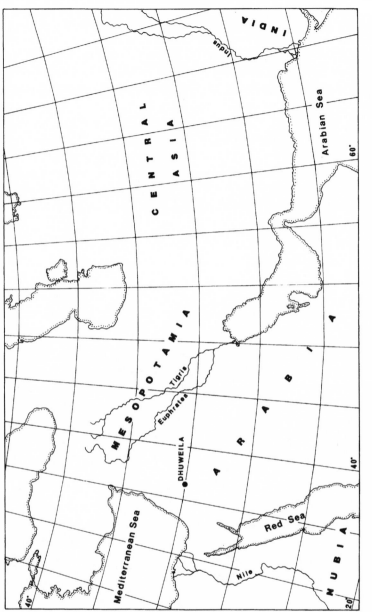

Figure 6.11. Map of Southwest Asia showing the location of Dhuweila. Reprinted with permission of Academic Press Limited, London.

dating to the period 3725 ± 725 B.C. Several pieces of lime plaster clearly showed fabric impressions, and small samples of the fabric itself adhered to some of the plaster. The impressions revealed a tabby or plain weave. The fabric on SEM analysis possessed neither the nodes exhibited by linen nor the scaly appearance exhibited by wool (Figure 6.10), the two fibers known to be in wide use during this period. Instead they had the smooth, twisting form of cotton (Figure 6.12). There has been some dispute as to whether cotton was first produced in North Africa (Egypt or Nubia) or in India, but the Arabian samples were much earlier than anything previously found in either place. There is little likelihood that cotton was actually produced in the area, because of climatic conditions. The investigators did not decide between the African and Indian sources, but they recognized both that the date of cotton manufacture had to be moved back and that important trade structures were in place at very early times.

Conservation scientists are actively engaged in cleaning, stabilizing, and reconstructing textiles. Samples may be cleaned aqueously with a mild detergent or organically with a dry-cleaning solvent. They may be fumigated with

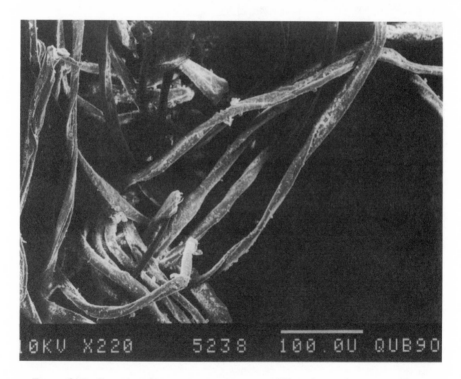

Figure 6.12. Scanning electron microscopic view of fibers from Dhuweila, that may be identified as cotton. Reprinted with permission of Academic Press Limited, London.

insecticides or fungicides to prevent further degradation. Textiles often are provided with a solid backing and on occasion are impregnated with polymers for stabilization. In the presence of metal objects, textiles become covered with metal oxides that accelerate decomposition (and may eventually form a pseudomorph). The conservationist may be able to treat the fabric with an aqueous solution of a material such as oxalic acid or citric acid that attaches itself chemically to iron, for example. The resulting iron oxalate or citrate can then be removed by rinsing with water. The procedure also works with iron stains on paper.

Skin, Hide, and Other Animal Parts

Long before people were twisting fibers, they were using the skins or hides of animals to serve as clothing and as coverings for any number of purposes, from musical instruments to war shields and shelter. Unprocessed hides, however, suffer from the dual problems of being putrescible (subject to natural and possibly unpleasant decomposition) and of becoming brittle. Skin is composed of the protein collagen, and hairs are composed of the protein keratin. Skin has two surfaces, the grain side that is exposed to the air and the flesh side that faces into the body.

Any chemical process for preserving skin is referred to as tanning, and any treated animal skin may be termed leather. The simplest tanning procedure is the application of oils or vegetable extracts to improve pliability and to make the surface impermeable to water. Such chemical processes probably date to Paleolithic times and could have used fish, egg yolk, or an infusion of oak bark. Prior to tanning, the skin often was processed by the removal of the hair (unhairing or depilation) in order to create a smooth surface for coverings or writing. Chemical removal of hair was superior to physical removal, such as scraping or shaving, which left an uneven surface. Skins were soaked in water to remove debris and dirt. Unhairing usually was effected by treatment with an alkaline, or basic, material, probably ashes originally but an aqueous lime solution for at least the last two thousand years. The ancient Egyptians and Mesopotamians used fermenting vegetable matter for unhairing. Many other steps have evolved in the processing of skins, which included fleshing (cleaning the flesh side of the pelt), scudding (scraping with a blunt knife), and splitting (separating the skin into two parts, a grain split and a flesh split). All these steps preceded the actual tanning with oils, minerals, and the like.

An alternative processing method for skins is felting, a combination of pressure, heating, and moisture that renders the hair irreversibly interlocked into a compressed mass. The earliest sample of felt may have been one found at Çatal Hüyük, dating to the sixth millennium B.C.

Skins have been used as writing surfaces for millennia but were first documented in Fourth Dynasty Egypt. This period (2613–2494 B.C.) marked the beginning of the Old Kingdom, during which Cheops (Khufu) was pharaoh and constructed the Great Pyramid. Papyrus later largely replaced skins for this purpose. About the second century B.C., parchment was invented in the Southwest Asian city of Pergamum, in present Iran, which in fact is the source of the word *parchment*. The invention supposedly was stimulated by the high price of papyrus, caused by an Egyptian boycott of the export of papyrus to cut off growth of the famous library at Pergamum during the reign of Eumenes II (197–159 B.C.).

To make parchment, skin is unhaired by liming but is not tanned. After chemical processing it is dried under tension on some sort of frame and then rubbed smooth with pumice (a rough volcanic stone). Parchment originally was sheepskin, but later goatskin was used as well. The term *vellum* is restricted to calf or lamb parchment. By Roman times both leather and parchment were stitched into rolls, but also sometimes into a flat tablet form, called a *codex* to distinguish it from a roll. Whereas the papyrus roll was the medium of choice in Greece and Republican Rome, the parchment codex was preferred in Imperial Rome and in later medieval times. Parchment was used for the famous Dead Sea Scrolls, found in Israel in the 1950s, dating to the second century A.D.

The species of animal from which skin was obtained may be determined from microscopic procedures. Characteristic grain patterns on skin vary with species and may be compared to standards to identify the species. Specimens also may be stained and the patterns of hair follicles examined for the same purpose. Dating may be carried out by carbon-14 methods or by determining the shrinkage temperature (the temperature at which soaked samples begin to shrink, older samples shrinking at lower temperatures).

Many famous documents have been rendered on parchment, but one of the most controversial is the Vinland Map. In 1957 a map was discovered bound into a thirteenth-century book now called *Tartar Relation*. The map, drawn on vellum, clearly shows the outlines of the Eurocentric world: Europe, North Africa, West Asia, Iceland, and even Greenland (Figure 6.13). It was the large island called Vinlanda Insula in the upper-left-hand corner, seeming to confirm the voyages of Leif Eriksson to America, that made the map at the time the most valuable in the Yale Map Collection. Exhaustive cartographical, paleographical, and philological analysis concluded that it dated from 1440 or earlier. Even the wormholes matched those of another document from this period. When Walter C. McCrone analyzed the ink, however, it proved to contain anatase, a variety of titanium dioxide unavailable until after 1920. In

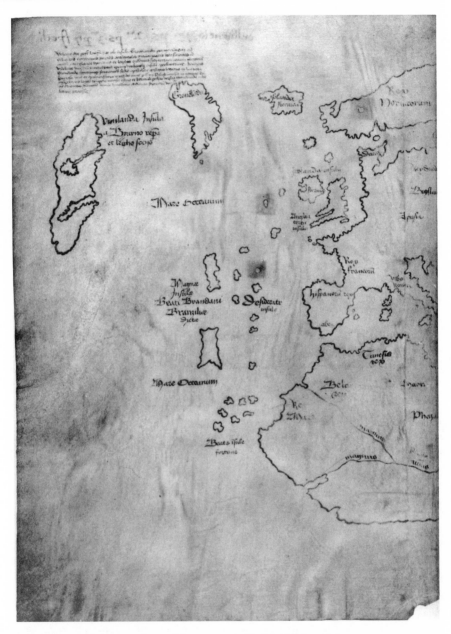

Figure 6.13. The western half of the Vinland Map.

1974 the Yale University Library declared that "the famous Vinland Map may be a forgery," a conclusion that by then had been generally accepted.

The tide began to turn almost immediately after Yale's acceptance of the map as a forgery. In the early 1970s Jacqueline S. Olin suggested that anatase could be produced naturally during the production of medieval iron-based inks. In the modern process, anatase is obtained by reacting titanium-containing iron minerals (usually ilmenite) with sulfuric acid, a process identical to medieval recipes for making inks. Any titanium present in the iron would become anatase. Titanium in fact is the ninth most common element in the earth's crust and is found in many common rocks and minerals. New studies of the Vinland Map by Thomas A. Cahill using the PIXE method (particle-induced X-ray emission) examined many more particles than McCrone and found much lower levels of titanium (ca. 0.01 percent) than did McCrone. Such levels have been found in other ancient inks, such as that used by Gutenberg. Thus the Vinland Map has been rehabilitated by science after being condemned by science. This is not to say that there is universal acceptance of the antiquity of the map. The final opinion must be based on cartographical and historical, as well as chemical, analysis.

Archaeologists, of course, exploit animal remains other than skin products, but soft tissues are rarely available. Bone, teeth, and ivory are common and provide a wealth of information. Much of such information is nonchemical in nature, such as species identification, frequency of appearance, techniques for butchery, evolution of domestication, and frequency of livestock diseases, from which conclusions about diet and environment are primarily drawn. Chemical analysis of human bone is examined in Chapter 8. Chemical analysis of animal bone often is carried out when the bone is artifactual in nature, such as a tool or carving.

Ivory artifacts in particular have been chemically examined. Like bone, ivory consists of the rigid inorganic matrix composed largely of the hard mineral hydroxyapatite (calcium hydroxyphosphate) and of the organic substance collagen, a protein in connective tissue. In an archaeological deposit, the organic portion tends to decay first, decomposing totally in as little as a few years or in more than ten thousand years. During burial the apatite portion takes up fluorine, uranium, manganese, iron, and other elements from the environment as contaminants. Thus bone that has been buried for some time tends to have decreased levels of nitrogen (a principal component of proteins) and increased levels of fluorine.

A group from the Institute of Fine Arts at New York University has studied the elemental composition of high-quality ivories from Southwest Asia. They found that modern elephant ivory from Africa contains about 5.5 percent nitrogen, 16.2 percent carbon, 53 percent ash (the inorganic residue after the organic part is burned off), and less than 0.1 percent fluorine. Ivory artifacts of reliable provenance from such Southwest Asian sites as Hasanlu (northwestern Iran, ninth century B.C.), Nimrud (Iraq, ninth and eighth centuries B.C.), and Megiddo (Israel, thirteenth and twelfth centuries B.C.) showed a consistent pattern of composition reflecting the ravages of time. The ash had risen to about 85 percent, indicating loss of much of the organic material. In agreement, the carbon content had dropped to 5 percent or less and the nitrogen content to 1 percent or less. The fluorine content was variable, from as low as 0.0 percent to as high as 1.45 percent.

This data set provided a reliable baseline for comparison with a group of ivories purchased by the Metropolitan Museum of Art in 1958 and attributed to the Iraqi town of Khorsabad (ancient Dûr Sarrukin, the unfinished capital of the Assyrian King Sargun from the eighth century B.C.). Figure 6.14 shows

Figure 6.14. Ivory plaque attributed to Khorsabad, Iraq. Reprinted with permission. © 1974 by the American Chemical Society.

one of these ivories. The chemical analysis fitted nicely into the pattern of ancient ivories, as the ash level was 84.9 percent, carbon was 4.4 percent, nitrogen was less than 0.4 percent, and fluorine was 0.06 percent. Subsequent comparison with actual specimens excavated from Khorsabad showed similarities, except for the nitrogen levels. The authors concluded that the pattern was consistent with an ancient origin, but the locale may not have been Khorsabad. Another sample of unknown provenance, but attributed to ancient Assyria, proved to have 60.7 percent ash, 10.8 percent carbon, 3.65 percent nitrogen, and 0.02 percent fluorine. These proportions correspond to a modern composition and suggest a forgery. It is more clear-cut to prove that a specific ivory is not ancient than to prove that one is. Forgers even have developed chemical techniques for inducing fluorine uptake and for extracting collagen, thereby simulating an ancient composition. The legitimate museum chemist often is just a step ahead of the forger.

Chemical analysis of ivory can produce information about its source as well as its age. Nikolaas van der Merwe, John Vogel, and their groups have developed a method based on the ratio of the stable isotopes of carbon (^{13}C and ^{12}C) to locate where elephant ivory came from. Although the method was developed largely to assist in the identification of illegal modern ivory, it should also apply to ancient ivory. Moreover, isotopic ratios cannot be altered by a clever forger with a chemist associate.

Wood and Other Plant Remains

The remains of plants are highly susceptible to decay, but seeds, pollen, phytoliths, and even soft tissue nonetheless may be analyzed by the paleobotanist in terms of climate, diet, and cultural development. The most useful plant remains are probably bulk wood, whose tree rings may give an absolute date, whose species may contribute to understanding climate, whose use may enhance our understanding of culture, and whose form and decorations may be aesthetically attractive. Fresh wood is composed mainly of three large organic molecules or polymers. About 40–45 percent of the weight is made up of cellulose, a string of ten thousand or more sugar units attached end to end. There are many different molecules that classify as sugars or carbohydrates, but only glucose is found in cellulose. A smaller polymer, called hemicellulose, makes up 20–30 percent of wood. Its molecules generally are shorter than those of cellulose, but they have a larger variety of sugar units, including mannose, galactose, and arabinose as well as glucose. Both of these large molecules are referred to as polysaccharides. (*Saccharide* is synonymous with *sugar*, and the molecules contain many of the sugar units.) The final component, also 20–30 percent, is lignin, a complex polymer of phenolic cinnamyl alcohols. A phenol contains the benzene ring attached directly to a hydroxy (-OH) group. The term *cinnamyl* refers to molecules with a carbon-carbon double bond (HC=CH) attached to a benzene ring. In this case a hydroxymethyl group (HOCH$_2$–) also is on the double bond, making the molecule an alcohol. Loss of water from reaction of either the phenolic or the alcohol hydroxy groups leads to polymerization and cross-linking (connections between chains that make the polymer extremely durable). The proportions of cellulose, hemicellulose, and lignin and the details of their structure vary among types of wood and influence the rate of decay. Insects, fungi, and bacteria all contribute to decay, as well as availability of air and water.

Wood may be characterized according to species by solubility tests, elemental composition (proportions of carbon, hydrogen, and nitrogen), carbon isotopic composition, nuclear magnetic resonance spectra of the solid wood, electron microscopy, oxidation with copper oxide, and mass spectrometry. All evidence indicates that the polysaccharides decay faster than the lignins, so that buried wood tends to be lignin-rich. Since the benzene rings of lignin have a higher proportion of carbon compared with polysaccharides, the process of decay causes an increase in carbon at the expense of hydrogen and oxygen. The ratios of these elements thus can be used as a measure of age.

Loss of organic components also lightens wood and decreases its mechanical strength, ultimately causing it to become brittle or spongy and to fall apart. Such a process occurs in soil, in air (as with standing totem poles), and in water. Waterlogged wood, when dried, undergoes cellular shrinkage and collapse that leads to warping, splitting, and flaking. Coatings or finishes such as polyvinyl acetate have been applied to dry wood in order to consolidate the surface, reduce flaking, and protect it from ultraviolet light, which accelerates chemical decomposition. For wood that was waterlogged, a more invasive treatment is usually necessary. Nonreactive polymers such as polyethylene glycol (PEG, -(OCH$_2$CH$_2$)$_n$-) are added slowly to the water. At low concentration levels the polymer moves into and stabilizes the cell structure, preventing it from shrinking on drying. The polymer does not truly react with the wood molecules but forms weak hydrogen bonds rather than full, or covalent, bonds. At higher concentrations PEG begins to fill permanent voids in the wood structure, such as cell lumens, further improving the mechanical properties. Treatment with PEG is considered to be reversible, but other bulking treatments involve actual covalent linkage with cellular molecules. Alternatively wood may be flooded with monomeric materials that can set into plastics, such as polymethyl methacrylate, by polymerization. Because these last techniques are irreversible, they are used only in extreme cases, and some say they should not be used at all.

Possibly the most daunting wood-conservation project mounted to date involves the Swedish warship the *Wasa*. This enormous vessel was poorly balanced on construction and sank in the Stockholm harbor on its maiden voyage in 1628. It was rediscovered in 1956 and has been in the process of salvage and conservation ever since. The pioneering work with PEG as a consolidant for archaeological wood had been carried out by Bertil Centerwall in Sweden, so the conservators of the *Wasa* were fortunate that the embryonic technology was available locally. After considerable deliberation, spraying of the *Wasa* hull was initiated on April 9, 1962, with a solution of PEG of various molecular weights and a little boric acid and borax as a fungicide to inhibit further deterioration. Eventually an automated spraying system was installed, and some modifications were made in the chemical makeup of the solution. Spraying continued into January 1979, for almost twenty years. Individual items such as sculptures were soaked separately in tanks. The success of the operation is still being monitored.

For historical purposes, one of the most important products made from plants has been surfaces for writing. By its nature, writing requires a substrate, which originally included ceramics, skins, or rock. The Code of Hammurabi (ca. 2030 B.C.) was carved in about four thousand lines on a block of feldspar 87 inches high and 74 inches around the base. Babylonian writers incised their messages in cuneiform, on clay tablets that were baked. Parchment was the highest level of development in the ancient world from these three materials. The Egyptians, however, developed the most successful ancient writing medium with papyrus. This plant *(Cyperus papyrus)* was something of a staff of life for the Egyptians even before 3000 B.C. They burned its woody part for fuel, manufactured furniture and utensils from it, ate its stalk as a vegetable, lashed its stalks together for boats, and plaited the fibrous cortex to form mats, rope, sails, and sandals. Referred to as "bulrushes," papyrus provided the legendary hiding place for the infant Moses, whose small boat was also made of papyrus.

To serve as a writing medium, beginning sometime before 3000 B.C., the Egyptians cut thin strips from the pithy center of the stalk, placed them side by side to form a sheet, and constructed another sheet with the strips laid at right angles to the first set. The material was moistened, pressed, and beaten to form a flat, cohesive mass, much in the fashion of making felt. The sticky contents of the pith served as an adhesive. Flour paste was used to connect individual sheets to form the papyrus roll or scroll. When dried, the material was polished with cedar-wood oil on one side, called the *recto* (the reverse side was the *verso*). Egypt practically had a monopoly on the papyrus reed. The Greeks imported it via the Syrian city of Byblus and gave it the name βυβλιου, from which the English word *Bible,* meaning "book," is derived. Papyrus became the chief writing material for Greece and Republican Rome, although wax tablets also were popular. The dry climate of Egypt resulted in the preservation of hundreds of thousands of papyrus documents, which have provided us with much of their classical literature as well as an extensive record of the daily life of authors of tax receipts, census records, religious practices, private letters, and birth, marriage, and death notices, exhumed from the garbage heaps of Egypt and elsewhere.

The manufacture of paper by the processing of vegetable fibers was invented by the Chinese by A.D. 200, but paper was used very little in the Western world for more than a thousand years. The Arabs introduced paper to Europe through their academies in Spain, but even in Southwest Asia it was not used extensively until the thirteenth century. By the fourteenth century it was well established in Europe, just in time for the invention of movable type by Gutenberg in the fifteenth century. Papyrus must still have been in use at the time of the introduction of paper, as the latter word is derived from the former.

Natural Products

Microscopic examination of most of the materials considered up to this point, from linen to leather, clearly shows a cell structure indicative of the biological source. Extensive processing of biological materials, either naturally or by human intervention, eventually leads to the total loss of the cell structure and a concentrating of certain organic molecules, which chemists call *natural products*. Thus the molecules that make up dyes, beer, wine, and bread (including ethanol) are natural products, as are the molecular markers in food residues on pottery. Dating back possibly to Paleolithic times, humans have used natural products for a variety of purposes, from therapeutics to glues. Analysis of these materials truly brings archaeology to the molecular level.

Geological processing of plant remains through high pressure and heat over millions of years leads to the formation of petroleum and coal products that are of critical importance as energy sources today. Both petroleum and coal are examples of extremely complex natural products, composed of a multitude of organic compounds, sometimes tied together in polymeric, high-molecular-weight forms. Occasionally, early humans found them to be useful or decorative. These earliest forms of natural products exploited by humans required little or no processing. They may have had some remanent biological structure, such as wood cells.

Petroleum products sometimes ooze from the earth as a tarry material called *bitumen*. The La Brea Tar Pits in Los Angeles have provided many examples of remains of ancient animals caught in the sticky substance. The creative Paleolithic humans of Southwest Asia may have been the first to use bitumen, probably as an adhesive. Two 36,000-year-old stone tools from Umm-el-Tlel in Syria contained bitumen residues that indicate the considerable antiquity of its use. Later examples have been found at Jarmo in Iraq, dating from about 7000 B.C. Figure 6.15 shows this site at the upper left, as well as many others. Processing of bitumen also began in this region. The addition of a mineral such as calcite ($CaCO_3$), silica (SiO_2), or occasionally gypsum ($CaSO_4$) creates *asphalt*, which was used extensively as an architectural material, first in the Zagros Mountains and later in the riverine cities of Mesopotamia. As bitumen and the minerals do not combine chemically, asphalt does not constitute a new chemical compound such as Egyptian blue. It is, like plaster, a shrewd and original chemical formulation that provides a useful application.

Figure 6.15. Sources of bitumen and asphalt in Southwest Asia. Reprinted with permission. © 1978 by the American Chemical Society.

Robert F. Marschner has analyzed bitumen products from Babylon, Jarmo, Susa, and other locations in Southwest Asia, as well as from the site of Mohenjo-Daro in Pakistan. His chemical analysis of asphalt led to distinctions between these sources. He determined the type of mineral (calcite or silica), the ratio of bitumen to mineral, and the percentage of sulfur in the bitumen component (5–10 percent). He found that lowland (Mesopotamian) and highland (Zagros) sources were distinctly different. At later sites, the asphalt had a significant component of vegetal matter, and its uses expanded beyond adhesive properties to include applications as sealants and protectants. Bitumen even became a trade item. In a recent excavation at Hacinebi Tepe in modern Turkey, where bitumen is not found naturally, archaeologist Gil Stein found several examples of bitumen, used during the period 3800–3100 B.C. as a sealant for ceramic containers. He attributed them to Mesopotamian sources and characterized Hacinebi Tepe as a trade outpost of the Mesopotamian Uruk culture.

Although we think of coal and petroleum as messy materials, a related substance known as *jet* has been used as a decorative gemstone since Upper Paleolithic times. Here geological processing has resulted in an organic stone with a hard, attractive appearance. The amulet in Figure 6.16 (left) in the shape of an insect larva dates to 15,000–10,000 B.C. and was found in the German Swabian Alps. By Roman times, jet medallions (Figure 6.16, right) had become common. Jet is an example of a carbonaceous material, a group that includes peat, lignite, bitumen, all forms of coal, and even graphite, in which hydrogen is entirely absent (graphite is pure carbon). We examined the NMR

Figure 6.16. Left, *amulet made of jet in the shape of an insect larva (38 mm long), 15,000–10,000 B.C., Upper Danube, Germany.* Right, *Roman jet betrothal medallion (40 x 49 mm), ca. fourth century A.D., York, England. Courtesy of the Yorkshire Museum.*

spectra of jet materials from England and Spain and were able to determine that jet is most closely akin to lignite and sub-bituminous coal. The spectra of English and Spanish coals were different enough that a distinction could be made between them. A sample of carved jet found in a Mayan tomb in Tipu, Belize, from the later sixteenth century was consistent by NMR analysis with a Spanish source.

Although bitumen requires no processing, the simple addition of a mineral to form asphalt produces a far more versatile product. The next step in processing organic raw materials may have been heating. Paleolithic people heated wood, bark, peat, coal, and probably many other natural materials, driving off volatile substances and altering the residual chemicals to produce a new composition called *pitch*. The resulting products varied somewhat, depending on the source substance. Later technology permitted the collection of the volatile substances given off by heating, a process we call distillation. The volatile products of distillation are called *tar*. When the tar is reheated to drive off new volatiles, the residue is called *tar pitch*. Tar and pitch have found an extraordinary range of uses over the millennia, including adhesives, sealants, coatings, medicines, ingredients of wine, incense, and even chewing gum (judging by tooth marks on some tar artifacts). Coal tar is still used in dandruff shampoos. The archaeological chemist is interested in determining the source material by molecular analysis. As with food residues, the analysis may consist of examining the whole substance by IR or NMR, or of separating it into molecular components by GC and identifying them by MS.

In an early example Richard Evershed and his group used IR, NMR, and GC/MS to identify the tars and pitches found on the Tudor vessel the *Mary Rose*, Henry VIII's flagship that sank in the Solent River in 1545. It was raised in 1981 and is considered to be of the same significance as the *Wasa*. Pitch or tar was found in numerous contexts, from unused barrels in the hold to impregnations on rope. The infrared spectra in Figure 6.17 show the similarity of a *Mary Rose* tar (B) to a modern Stockholm tar (C) made from pine wood. The bottom spectrum (D) is of a spruce wood tar and is clearly different, particularly in the region from 1400 to 1800 cm^{-1}. The top spectrum (A) is of pitch found in the 1983 excavation of an Etruscan vessel that had sunk about 600 B.C. off the island of Giglio, 80 miles northwest of Rome. The similarity to Stockholm tar also suggests a pine source. Evershed carried out a similar analysis of an adhesive used on a Roman jar found in the River Nene in Northamptonshire. The jar was intact except for a circular piece that had been glued on in antiquity with a tar. By GC/MS analysis, Evershed and his co-workers proved that the source of the tar was birch bark. The key molecular markers were betulin, lupeol, and lupenone. The terpenes characteristic of pine tar and the alkanes characteristic of bitumen were absent.

How did ancient people create tar without modern distillation equipment? Medieval alchemists used glass or metal devices called retorts, which seem to crop up in any depiction of an early chemistry laboratory: They have round bottoms with a long sidearm leading off to a point from which the distillate drips. The Latin verb *stillare* is translated as both "to distill" and "to drip." More ancient equipment may have resembled the medieval device shown in Figure 6.18. Fuel surrounded a ceramic container with holes on the bottom, and the tar dripped out of the bottom. This process may not have been true distillation, which requires the product to vaporize and condense. No archaeological evidence on how tar was produced predates medieval practices, yet tar has been produced certainly for five thousand years and possibly for ten thousand.

Pine tar can be made not only from wood or bark but also, and commonly, from the sticky material called *resin* that oozes out of a tree from wounds. Unheated and otherwise unmodified resins were used in the ancient world as adhesives and linings and later as coatings or binding media for paintings. John S. Mills and Raymond White have made an extensive study of these

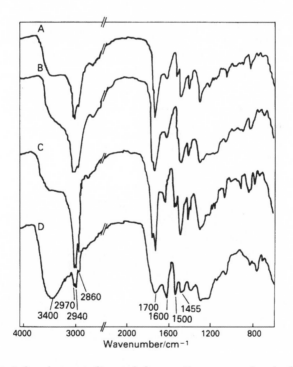

Figure 6.17. Infrared spectra of A, *pitch from an Etruscan vessel sunk off the Italian island of Giglio, ca. 800 B.C.;* B, *tar from the Mary Rose;* C, *modern Stockholm tar made from pine wood;* D, *Norway spruce tar. Reproduced with permission of the Royal Society of Chemistry.*

materials, primarily by GC/MS. Natural resins are composed almost entirely of a class of compounds called *terpenes,* which are made up of five-carbon blocks. Monoterpenes have 10 carbons, sesquiterpenes have 15, diterpenes have 20, and triterpenes have 30. When the substance contains only monoterpenes, it is volatile, odoriferous, and called an *oleoresin.* Interestingly, diterpenes and triterpenes are rarely found together in a resin. Thus conifers (plants bearing cones) contain diterpenes such as abietic acid, agathic acid, pimaric acid, and labdane. Unprocessed pine resin has sometimes been called common turpentine. (The term *turpentine* is used for resin from other sources as well.) On destructive distillation, pine resin gives oil of turpentine as the volatiles (hence a tar), and the residue is called rosin or colophony (hence a pitch). Triterpenes are found in large tropical trees such as dammar and the genus *Pistacia,* which is the source of mastic. When resins have a large carbohydrate (sugar) component, they are called *gum resins,* which include the legendary frankincense and myrrh, which were used for perfume and medicine.

The method of choice for identification of ancient resins has normally been GC/MS, since they are made up of numerous molecular components. The group at the University of Illinois identified the preservative resins used on their mummy. They found several derivatives of abietic acid and concluded that the material was a pine resin. They also found bitumen, which was identified by a series of hydrocarbons. Because the carbon in bitumen was produced

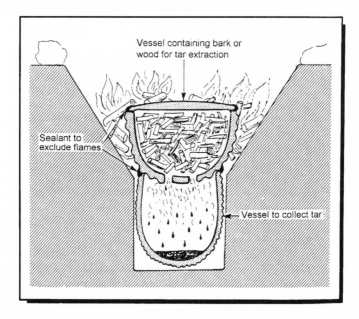

Figure 6.18. Reconstruction of a medieval apparatus for producing tar.

by plants millions of years ago and small amounts of bitumen contaminated the wrappings, the radiocarbon date was off by a few hundred years. Curt W. Beck used IR and GC/MS for the study of residues and linings of amphorae found in the harbor of ancient Carthage (Tunisia, from Punic to Byzantine times, fourth century B.C. to seventh century A.D.). He found diterpenes related to abietic acid in more than 80 percent of the cases and concluded that the material was pine resin. It is difficult to tell whether the material was unprocessed resin or pine tar. In his study of the currently oldest sample of wine (5400–5000 B.C.), Patrick McGovern found samples of the oleoresin from the terebinth tree. This resin probably served to inhibit bacterial growth and to impart a particular aroma to the wine.

As resin ages, chemical changes occur that result in a hard and very stable material that often is called *amber.* Fossilized resin is typically 20 to 50 million years old but is known with an age of more than 200 million years. Because of its attractive appearance and ease of working, amber has been a valued raw material for carved products since Paleolithic times. Its widespread occurrence in northern Europe was the result of the presence of a vast forest of resin-producing trees during the Tertiary Period. Glaciation moved the amber deposits to a broader area, which is represented by the shaded portion of the map of Europe in Figure 6.19. This type of amber often is called succinite from the Latin word for amber, *succinum,* but other types of fossil resin have been found throughout Europe (filled circles in Figure 6.19), as well as in many other places in the world, such as New Zealand, Greenland, the Caribbean, North America, and China. Amber has been found in numerous archaeological contexts. On finding it in the shaft graves of Mycenae in the late nineteenth century, far from the sources around the Baltic Sea, Heinrich Schliemann commented, "It will, of course, remain a secret to us whether this amber is derived from the coast of the Baltic or from Italy." The movement of amber from northern to southern Europe in fact probably created the earliest trade routes, from 5000 B.C.

The pessimistic prediction of Schliemann has been disproved by a number of different techniques. Curt Beck pioneered the use of IR to sort Baltic from non-Baltic amber, using the characteristic absorption pattern from vibrational stretching of carbon–oxygen bonds in the region 1110 to 1250 cm^{-1}. Baltic amber gives a flat absorption, as seen at the right of Figure 6.20, shared by no other source of fossil resin. We have used carbon-13 NMR to establish distinctions on a worldwide basis, and several groups have used GC/MS to achieve similar aims. Beck analyzed numerous samples from archaeological sites in Greece, including Mycenae, and found that the vast majority were of Baltic origin. We established by NMR that amber found in a postconquest Maya tomb in Belize also was Baltic, presumably brought by the Spanish conquerors.

When a natural resin is dissolved in a solvent, the solution is often referred to as *varnish*. It is to be distinguished from *shellac*, a solution of deposits (called lac) from the insect *Coccus lacca* from India, and from *lacquer*. Although the term *lacquer* sometimes refers to any coating, such as varnish, that dries solely by solvent evaporation to leave a protective coating, true lacquer is a highly complex material developed in China and Japan, dating to the Shang Dynasty. The sap of a shrubby tree, *Rhus vernicifera* or *Toxicodendron verniciflua*, is extracted by tapping the bark in the same way that maple sugar or natural rubber is gathered. This sap, called *urushi*, is extremely irritating to the skin, as it contains many of the same components as poison ivy *(Toxicodendron radicans)*. It is composed of 20–25 percent water, 65–70 percent a mixture of phenolic materials called catechols, about 10 percent gummy substances, and less than 1 percent the enzyme laccase. A layer of the material is

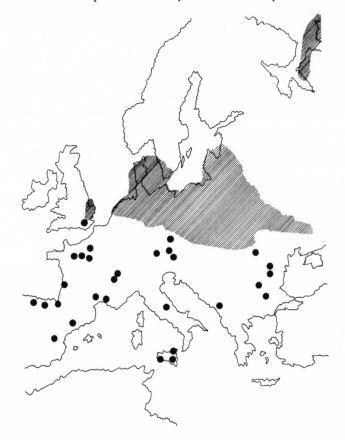

Figure 6.19. Map of Europe showing (shaded areas) *the source of succinite or Baltic amber and* (filled circles) *the sources of non-Baltic amber. Courtesy of Marcel Dekker, Inc.*

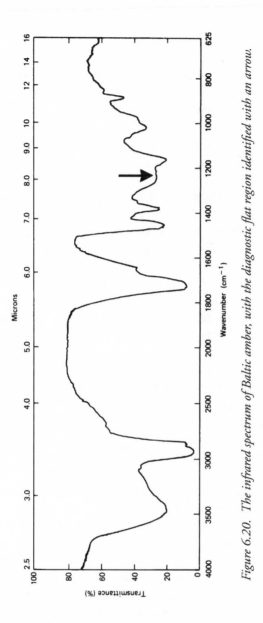

Figure 6.20. The infrared spectrum of Baltic amber, with the diagnostic flat region identified with an arrow.

applied to the surface of the object, which then is rubbed and polished until mirror smooth. The process is repeated up to thirty times, sometimes with pigments added to provide designs. The drying of each layer is catalyzed by the enzyme and involves reaction with the atmosphere. The resulting molecular structure contains an intricate network of bonds connecting the oxygen atoms with carbon rings and chains. Sometimes carbohydrate molecules from the gummy part are interwoven. The highly cross-linked material is extremely hard and impervious to chemical attack.

A. Burmeister has studied lacquer by mass spectrometry, and we have studied it by NMR. The structure of the final product generally mirrors the starting urushiol. Burmeister observed compositional differences that evolved over time, which may reflect changes in techniques of harvesting, preparation, or coating. Although lacquer boxes and trays are familiar to many people, lacquer also has been used to coat saddles, sword scabbards, pillars on temples, and whole pieces of furniture. Technology has not yet been able to say a lot about where a particular lacquer object was made or what techniques were used.

Glue is another word used loosely for adhesives, but it should be restricted to materials of animal origin. Collagen, the protein that makes up skin, hide, muscle, and the organic part of bone, is normally insoluble in water. Under high temperature and pressure, water can hydrolyze collagen, and the proteins are broken down to smaller units called polypeptides. The hydrolyzed material thickens on cooling to form glue. It may be recognized by characteristic NMR or IR patterns.

Like resins, *gums* are exuded from plants that have been damaged. Gums, however, are largely composed of water-soluble carbohydrates. They have been used as the medium for watercolor paint, as an adhesive, and as food. Specific gums generally are identified by their plant source: gum arabic *(Acacia senegal),* cherry gum *(Prunus cerasus)*, and so on. John W. Twilley has described methods for analyzing gums in terms of the constituent sugars, but there have not been extensive scientific studies of ancient gums. Gum, glue, resins, and other materials also are used as additives, called *size* or *sizing,* to paper, textiles, leather, and even metal to fill in pores and smooth the surface.

Waxes are composed of long chain molecules, particularly the chemical combination of organic acids and alcohols called esters. They can come from either the plant or the animal kingdom. Raymond White developed simple GC methods for distinguishing many types of wax. Figure 6.21 shows gas chromatograms for several common examples. The series of peaks for paraffin wax represents the alkane hydrocarbon family. Beeswax has two groups of peaks—several hydrocarbons on the left and a family of esters on the right. These chromatograms may be used as fingerprints for a particular wax or

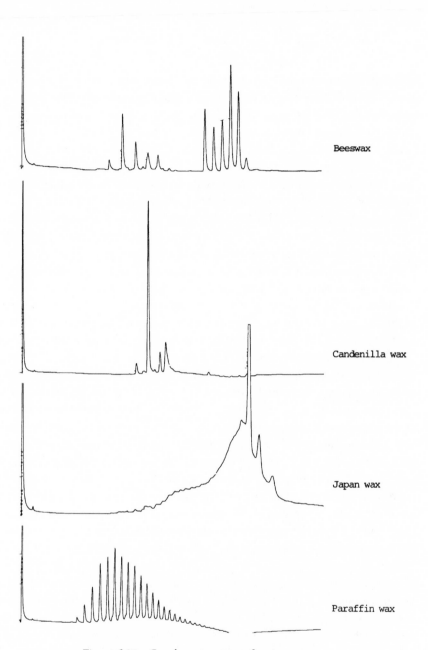

Figure 6.21. Gas chromatograms of various waxes.

combination of waxes. J. Glastrup identified the wax from a painting called *Cornfield with Three Haystacks* as consisting of a mixture of beeswax, paraffin wax, and the resin dammar. A group led by M. Cassar found by NMR that the wax used to seal royal documents of King Stephen (1135–1154) and King John (1199–1216) of England was composed almost entirely of beeswax.

The last category of natural products to be considered includes *oils* and *fats*. These materials differ only in that oils are liquids and fats are solids. They are formed by the attachment of long chain organic acids ("fatty acids") to the molecule glycerol, $HOCH_2CH(OH)CH_2OH$. Waxes, fats, and oils together often are called *lipids*. Oils may be extracted from plant materials such as olives, sunflower seeds, cotton seeds, coconuts, linseed, and castor; and from animal products such as cows' milk, chicken eggs, and pig flesh. Glycerol is linked to three fatty acids by ester linkages, which often are hydrolyzed under conditions of burial. Glycerol generally dissolves in ground water and disappears, leaving a collection of fatty acids for analysis. The presence of specific fatty acids can be used as markers for specific oils. Olive oil was suggested to be the residue in Roman amphorae and lamps because of the presence of palmitic, oleic, and linoleic acids, which are the known main components of olive oil. A group from the University of Keele in England tentatively identified the contents of barrels from a submerged Basque shipwreck (the *San Juan*) off Labrador as whale oil, from the presence of myristic, palmitic, gadoleic, and cetoleic acids. Large deposits of animal fat, on burial in wet ground, turn into a solid waxy material, composed of fatty acids, called adipocere or bog butter (in peat bogs). Certain oils, called drying oils, have double bonds in their long chains that can undergo polymerization on exposure to the atmosphere (see Chapter 4). This phenomenon is useful for coatings. Common drying oils include linseed oil, known to the Egyptians, and poppyseed oil, known to the Greeks and Romans. Oil varnishes contain mixtures of resins and drying oils.

As early as the third millennium B.C., the Sumerians were able to produce liquid soap by the action of an alkali (a base containing alkali or alkaline earth metals) on plant or animal fats. Alkali liberates the fatty acids as water-soluble salts, which are very effective in dissolving grease and other food stains into water. The chemistry here is not simple and may have developed from the use of ashes, which are alkaline, to cleanse skin or clothes by a combination of abrasion and chemical action. Someone had to discover that the solution from combining ash and grease could by itself serve as a cleaning agent, because ash (alkali) and grease (fat) provide the two components of soap. It was not, however, until about A.D. 800, perhaps in the Gaulish town of Savona, that an aqueous solution of soap was treated with metal salts to precipitate a solid and create the material we generally associate with soap.

The world of organic chemistry is enormous, but it has only begun to impinge on archaeology. The fragile nature of organic materials such as clothing and wood make them rare and valuable as artifacts, so that chemical analysis has been secondary to preservation. Even robust materials like amber and jet were not popular in analytical circles until recently because of their molecular complexity. As a result, chemists directed their attention toward the inorganic world of less fragile materials with a more straightforward chemical composition. This point of view now has been abandoned, largely because the analytical tools of organic chemistry have progressed. Very small amounts of materials can be separated by chromatography into distinct organic components, whose structures can be deduced by spectroscopic methods. Materials such as food, drink, adhesives, and oils that were never previously considered to be traditional archaeological artifacts nonetheless have been found to leave chemical residues that now may be analyzed to help understand the culture of early peoples. The human body, too, is a library of organic compounds that leave residues, whose analysis is the subject of Chapter 8.

7

METALS

Native Metals

For hundreds of thousands of years, our species and its immediate predecessors (hominids) chipped, cut, and shaped stone, bone, and wood. The experimenters among them must have tested many conceivable types of available raw material over the millennia. Many familiar stones, such as those considered in Chapter 1, are composed of metals bound to nonmetallic groupings such as carbonates or silicates, or to nonmetallic atoms such as oxygen (oxides) or sulfur (sulfides). Very rarely, stones are composed of pure, unbound, or *native metal,* such as copper, iron, or gold. When early humans applied their chipping techniques to these metallic stones, however, they obtained a different response. The stones did not chip or flake like flint or obsidian but stretched and bent into new shapes. Only when their plasticity was recognized as an advantage did the age of metals begin to dawn. The durability of metal was clearly superior to that of wood. A deformed metal implement could be repaired, in contrast to a damaged one made of nonmetallic stone. The malleability of metal enabled it to be hammered into new shapes, such as flat sheets. Its ductility allowed it to be pulled into an elongated form. Its attractive colors and luster led to the creation of objects meant to be appreciated solely for their beauty.

The Neolithic people who first discovered these advantages of metal lived ten or twelve millennia ago, almost certainly in Southwest Asia. They lived in small villages and subsisted by farming, hunting, and fishing. In addition to making tools from stone, wood, and bone, they formed clay into pottery and wove fibers into fabrics. The discovery of the properties of metals, along with the development of farming and the domestication of animals, provided the necessary ingredients for the rise of urban civilizations, whose place in time eventually was called the Bronze Age, after these new materials. The exploitation of metals required the enhancement of existing trade routes by which obsidian, flint, amber, turquoise, and bitumen were procured. The new crafts, but particularly the working of metal, required specialists and hence encouraged social stratification. The earliest metal products included not only tools and ornaments but also weapons of war: axes, knives, arrowheads, and maceheads. One group of people could impose their way of life on another and thereby create a larger social entity.

Only a few metals occur in nature unbound to other elements—that is, in a native state. These include gold, silver, iron, and most importantly copper. Gold and silver are called noble metals because of their low affinity for bonding to other elements. Native iron occurs most commonly in meteorites. Native copper is found throughout the world but was first exploited in Southwest Asia and later around the American Great Lakes. Metals, including tin, lead, zinc, terrestrial iron, and the vast majority of copper, normally are combined with oxygen from the atmosphere or sulfur from the earth as oxide and sulfide minerals.

Once the properties of native copper were appreciated, it probably was worked by placing a nugget on a stone and pounding it with a harder stone into a new shape. Such cold-worked copper can be recognized today by polishing its surface and examining it with reflecting light under an optical microscope. Figure 7.1a shows the microstructure of native copper from Talmessi, Iran, with 50x magnification, and Figure 7.1b shows a sample from the same mine that had been hammered by the metallurgist Cyril Stanley Smith. The large grains of the unworked metal have become distorted and reduced in size, in this case by 82 percent.

Flint and other nonmetallic stones had been heated in fire, or annealed, since Paleolithic times, in order to improve their fracturing properties. Although native copper, when treated in such a fashion, did not fracture, it could be worked more easily. Cold-working can cause the metal to become hard and brittle, but heating restores its malleability and ductility. Thus copper could be worked, annealed, and then cold-worked again until the desired shape was achieved. An improvement involved working the solid copper while it was still hot—that is, by *forging*. The Old Copper Culture of North America (3000–1000 B.C.) was well versed in annealing methods, as Figure 7.1c shows. The microstructure of this spearhead lacks the deformed grains of cold-worked copper seen in Figure 7.1b. Instead the grains are large, well formed, and sometimes split into twins, all characteristic of annealing above 700°C, according to Smith. In North America, copper working was restricted to heating below the melting point, which is 1083°C. Higher temperatures were never achieved in the ceramic technology of the region. As a result, copper was not liquified, and the advantages of casting were not discovered by the Old Copper Culture. The sophistication of their products, however, is illustrated by the flat copper ax from the Hopewell Culture of Mound City, Ohio (Figure 7.2).

The sequence of cold-working and annealing had been discovered earlier in Southwest Asia, where native copper was found in reasonable abundance. Worked copper artifacts have been dated to 8500 B.C. in Iraq, to the eighth millennium in Anatolia, and to the seventh millennium in Mesopotamia. The metalworkers in this region, however, knew how to heat copper above its melting point, presumably during the process of annealing, and were able to produce the molten metal. The concept of shaping a metallic object by *casting* was then developed. It is very difficult to distinguish an annealed artifact from one that was cast. The grain sizes of unmelted native copper generally are larger, and the metal tends to be free of inclusions such as copper oxide or silver. In the native state, copper oxide should be found only on the metal sur-

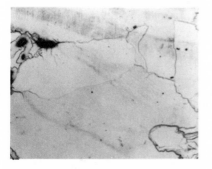

(a)

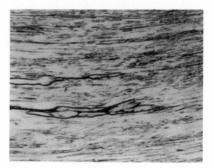

(b)

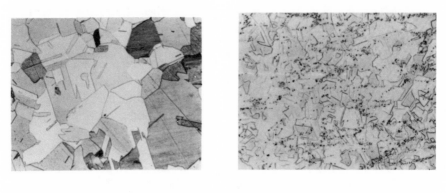

(c)

(d)

Figure 7.1. a, *magnification (50x) of native copper from Talmessi, Iran.* b, *magnification (50x) of native copper from Talmessi, Iran, that had been hammered cold to an 82 percent reduction in thickness.* c, *magnification (50x) of an annealed copper spearhead from the Old Copper Culture.* d, *magnification (200x) of a cast copper tool from Sialk, Iran.*

face, but this material is distributed throughout the object when it is melted. Figure 7.1d shows a 200x magnification of a tool from Sialk, Iran (fifth millennium B.C.), in which a network of particles of copper oxide is visible throughout the object as small black dots, a presumed indicator of casting. The development of casting native copper was an important and necessary step toward smelting.

Chapter 1 illustrated how patterns of trace elements are very useful in determining sources of flint, obsidian, turquoise, and other stones. This same method can be applied to copper sources, provided that the native copper had not been melted, since this process could mix samples from different sources. George Rapp, Jr., has been surveying the trace element composition of native copper. His early work centered on sources around the American Great Lakes, but his analyses now have run into the thousands and include sites from all over the world. Based on the levels of twenty-seven trace elements, he has constructed several statistical clusters. Figure 7.3 shows how samples from Illinois may be distinguished from samples from the Snake River of Minnesota (the coordinates are statistical combinations of elements). Using these sorts of tests, Sharon Goad and John Noakes demonstrated that copper artifacts found at the Copena site in Alabama had come from the Great Lakes, more than a

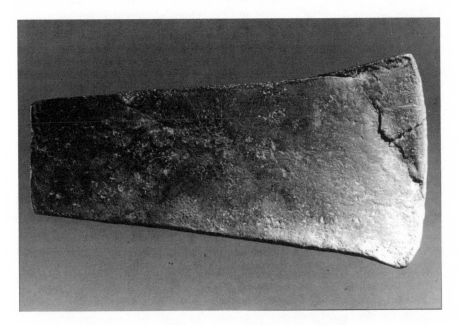

Figure 7.2. Native copper ax from the Hopewell Culture, Mound City, Ohio, with maximum length 18.4 cm and weight 1.024 kg. British Museum. Courtesy of B. Melton and the British Museum.

thousand miles away. Ronald G. V. Hancock and his group compared early American copper with European trade copper that natives began to use as early as the sixteenth century. They were able to distinguish American and European copper artifacts of this period by differences in the trace levels of gold (Au), arsenic (As), antimony (Sb), silver (Ag), and nickel (Ni), as illustrated by the excellent separation in Figure 7.4 based on just two of these elements.

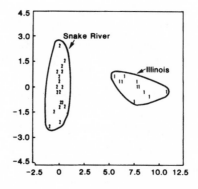

Figure 7.3. Separation of native copper from the Snake River, Minnesota, and from Illinois sites, as a plot of two principal components of elemental concentrations. Reprinted with permission. © 1984 by American Chemical Society.

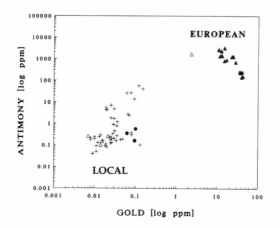

Figure 7.4. Separation of American local copper and European trade copper from the sixteenth and seventeenth centuries, as a plot of antimony versus gold concentrations. Triangles are archaeological materials, filled circles are modern copper wire, and plus signs are geological native copper.

Smelted Copper

In Southwest Asia, possibly before 6000 B.C., the workers of metal moved from using native copper to extracting it from its oxide and sulfide ores—that is, to *smelting*. This was an intellectual and technical landmark in the history of the human race, as these people had to be able to conceive of the presence of metallic copper in dull or differently colored rocks and then to alter the rocks chemically. How it happened is not certain, but various scenarios have been suggested. Cyril Smith and Paul Craddock imagine the process as evolving directly from the casting of native copper. Native copper mixed with small amounts of attached, undesired rocky material, made of silicates or iron oxides (collectively called *gangue*), but containing copper oxides, was heated with the intention of separating the gangue from the pure metal. Carbon from the fuel stripped oxygen from the copper oxide (the process of *reduction*) and produced additional metallic copper. The silicates and iron oxides also would have melted and turned into a glassy by-product called *slag*. The perceptive metalworker realized that more copper was produced than went into the fire and reasoned that it came from the rock.

An alternative scenario is more closely related to pottery production. The relationship between metallic copper and the green, copper-containing pigment malachite might have been observed when native copper naturally turned green over time or when malachite that was used as a pigment on a ceramic surface was heated in a reducing fire and produced free copper. Temperatures of around 1100°C are required for this process to occur, but this temperature is reasonable for a charcoal fire.

By either scenario, knowledge of the properties of melted native copper led to the recognition that copper could be produced by processing certain other rocks, or ores. Because the Old Copper Culture in North America never progressed to melting copper, they never discovered smelting.

The earliest physical evidence for smelting is some slaggy material found in association with copper beads at Çatal Hüyük in Anatolia, dating to 7000–6000 B.C. It is not clear if the slag came from true smelting or was a by-product of native copper casting. It is difficult or impossible to determine today if a metal artifact was produced from native or from smelted copper by chemical or microscopic analysis, either of the metal or of associated slag. James Muhly has suggested that one of the best indicators of the discovery of smelting was the increased scale of copper production. In the absence of smelting it was a cottage industry, but with smelting, vastly larger amounts of copper could be produced. An alternative indicator is the presence of elemental lead, which does not occur in nature as the free metal. Metallic lead would have been produced when lead ore had been subjected accidentally to the conditions for smelting copper. The earliest lead artifacts may be a bracelet from the sixth-millennium sites of Yarim Tepe in Mesopotamia and a lead bead from the contemporaneous site of Jarmo in the Zagros Mountains.

For three thousand years, copper metallurgy developed very slowly, leading to large-scale production of copper only during the fourth millennium B.C., sometimes called the Chalcolithic or Eneolithic Period, as a transition between the Neolithic Period and the full Bronze Age. The earliest smelting sites probably were open hearths or crucibles, into which the ore and the fuel were mixed. Figure 7.5 illustrates a clay hearth from the Chalcolithic site of Los Millares in Spain. Associated with it were copper ore, droplets of metallic copper, and fragments of other dishes or crucibles. Simple open hearths may not have been hot enough to form a free-flowing slag. The earliest method for increasing the temperature may have been the use of blowpipes. Clay pipes or tuyeres leading directly into the interior of the crucible, however, were not found.

Probably during the fourth millennium, fluxes were added to the ore to reduce its melting point and to create free-flowing slag. The flux could have been an alkali, silicate, or iron oxide and might have been present in the ore

Figure 7.5. Clay hearth from the Chalcolithic site of Los Millares, Spain, used for smelting copper.

itself. The presence of a tuyere (Figure 7.6) created a true furnace, whose temperature was increased by the flow of air. Eventually furnaces were constructed from which the slag could run off, or be tapped. As a result, lower-grade ores could be used, and larger amounts of slag were produced. The glassy slag offered properties of its own and contributed to the discovery of glaze and glass.

Craddock has pointed out that fluxed furnaces also were capable of reducing iron oxides in the gangue to metallic iron. Thus copper products contained significant iron impurities only when fluxes and furnaces had been introduced. That is why the material collecting at the bottom of the nonslag-tapping furnace of Figure 7.6 is called ferruginous copper in the drawing. Craddock found that copper smelted by primitive crucible methods contained only about 0.03 percent iron, but that copper from more advanced fluxing methods had about 0.3 percent, with great variability. The earlier group included predynastic Egypt, early Cycladic Greece, Bronze Age Britain, and Chalcolithic Spain, and the later group included Second Dynasty Egypt, Minoan Greece, Iron Age Britain, and Phoenician Spain.

There has been persistent evidence of an independent discovery or at least exploitation of copper smelting in southeastern Europe. Certainly by the fifth millennium B.C. both open-pit and shaft mines were in operation in Yugoslavia and Bulgaria. Remains of actual smelting, though, are sparser. Slag deposits in this area have been found at several Vinca Culture sites from as early as 5200 B.C., including Anzabegovo, Sekevac, and Gornja Tuzla. By the mid-fourth millennium, slag from Novacka Cuprija showed evidence of copper mixed with arsenic. These observations, made mostly since 1979, suggest a robust early copper metallurgy deep inside Europe.

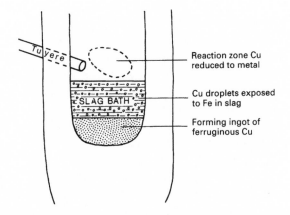

Figure 7.6. Diagram of an early smelting apparatus. Courtesy of B.R. Craddock.

The Discovery of Bronze

Experimentation with raw materials also led to metallic mixes, or *alloys*. Arsenical copper, sometimes called arsenical bronze, was the earliest such material. Copper containing about 2 percent of arsenic (As) melts more easily, is harder, and is more soundly cast than pure copper. Like gold, copper in the pure state is soft and bendable in a thin layer. Copper is hardened by cold hammering and by annealing, but only up to a point. Introduction of the second metal provided a much better tool or weapon. The advantages of arsenical copper may have been recognized when a copper ore naturally containing arsenic was used (such as olivenite, Cu_2AsO_4OH) or when an arsenic ore (such as the pigment orpiment, As_2S_3) was added as a flux. These materials are not widely available, so that even in Chalcolithic times extensive trade routes had to be developed. Arsenical copper ores, however, were widely available in northwestern South America and led to a long and fruitful production of arsenical copper.

Use of arsenic was quite uneven during Chalcolithic and Early Bronze Age times. A study by Craddock and his colleagues of Chalcolithic sites in southeastern Spain (Millarian culture, third millennium B.C.) found an average of 2.3 percent As (arsenic) in twenty-seven objects from Los Millares and 1.7 percent As from sixteen objects from El Malagón. The highest As content, up to 6 percent, was found in blades, for which the additional hardness would be useful. The tin (Sn) content of all these objects was less than 0.4 percent. A study by D. Britton of copper artifacts from the Early Bronze Age Wessex Culture in southern Britain (third millennium) found As above 0.2 percent in 83 out of 107 objects. The highest levels, above 2 percent, were found in seven daggers—again, materials that could benefit from the additional hardness. Unlike the Millarian objects, almost all the Wessex objects also contained varying but significant amounts of tin.

In the Old World, tin emerged as the preferred alloying element, to form tin bronze (usually referred to simply as bronze), during the millennium of experimentation preceding the Early Bronze Age, about 4000 to 3000 B.C. The advantages of tin bronze were much the same as those of arsenical copper—ease of casting, especially in closed molds, and a harder object that could be toughened by working better than could pure copper. The earlier objects had only small amounts of tin, 1–4 percent, sometimes mixed with arsenic, as at the Wessex Culture sites. A tin content of about 10 percent was settled on as optimal only after long experimentation.

During the full extent of the Bronze Age, from about 3000 until about 1200 B.C., gradual improvements were made in mining and smelting. Lower grades of ore were improved first by *beneficiation,* by which crushing, grinding, washing, and sorting led to higher-grade ingredients (or charge) for the furnace. A wider variety of ores were processed, including sulfides as well as oxides. Sulfides normally are found deeper in the earth than oxides and hence are more difficult to mine. Initially, sulfide ores were converted (or *roasted*) to oxides by treatment with oxygen before smelting. The process could involve nothing more than heating the crushed ore in the open, to bring about the chemical conversion of Equation 7.1. Oxide ores, or the roasted sulfide ores, were then smelted by the process of Equation 7.2, in which the oxide was reduced to metallic copper with carbon monoxide. This gas was formed in the furnace from the reaction of charcoal or another fuel with oxygen or with copper oxide. More advanced techniques involved reacting sulfide ores, which were more abundant than oxide ores, directly with oxygen to convert copper sulfides to copper metal. Such a process involves oxidative rather than reductive smelting (oxygen is added to copper initially rather than removed from it).

$$Cu_2S + 2O_2 \longrightarrow 2CuO + SO_2 \qquad (7.1)$$

$$CuO + CO \longrightarrow Cu + CO_2 \qquad (7.2)$$

To make bronze, tin could be included in the smelting mix as the oxide (cassiterite, SnO_2) or the sulfide (stannite, $SnCu_2FeS_4$) in the presence of charcoal (to remove the oxygen or sulfur by reduction), or directly as pure tin metal. In later stages of the Bronze Age, lead was included in the charge to give rise to leaded bronze. A variety of fluxes were added to improve slag formation, although they had to be selected to be compatible with the specific mix of metals. Furnaces were improved by the development of shafts or chimneys to enhance air flow, and the use of a variety of bellows increased the temperature and direct oxygen into the charge. Modern re-creations of ancient furnaces have enabled furnace temperatures to be determined. For example, the Bronze Age furnaces at the large Timna mines in the Sinai appear to have operated at 1100–1200°C. Refining could lead to further purification. Fire refining involved melting the metal in an open crucible with stirring. Volatile components such as arsenic evaporated, and more reactive components such as iron were oxidized to form a surface layer that could be skimmed off.

The molten metal from the smelting furnace was poured into a variety of molds. The earliest molds were sculpted directly from stone to produce axe- and spearheads (Figure 7.7, from Ballyglisheen in County Carlow, Ireland). Clay and permanent bronze molds were used in the Middle and Late Bronze Age. Two-part molds facilitated disassembly, and use of the *lost wax method* permitted enhanced detail. By this procedure, an object was sculpted in hard

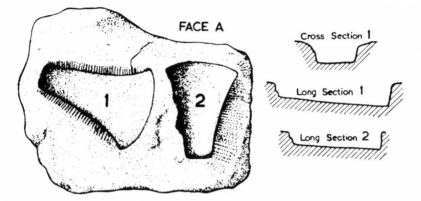

Figure 7.7. Stone molds for bronze axes, from Ballyglisheen, County Carlow, Ireland.

wax, and a clay form was pressed around it. The process of firing hardened the clay and melted the wax, leaving an interior in the desired shape. Pouring molten alloy into the clay mold then produced an object with the same shape as the original sculpted wax. The initial wax piece could constitute the entire figure, or it could be sculpted over an earthenware core. In the latter case molten bronze was poured between the core and the outer mold, so that the result was a shell rather than a solid, hence lighter and less wasteful of metal. Bronze often was first cast as ingots, or pigs, for trade, storage, or further treatment.

Bronzes with substantial amounts of tin have been found in the royal graves of Ur in Mesopotamia from about 2800 B.C., in Fourth Dynasty Egypt (2600 B.C.), in Troy III (2200 B.C.) on the Anatolian coast, and in Middle Minoan III sites (1700 B.C.) on Crete. This tradition eventually led to the delicate virtuosity of the fifth-century B.C. Greek artists who excelled in bronze as well as stone and pottery (Figure 7.8).

Bronze in Central Asia, East Asia, and Africa

From its Old World source in Southwest Asia, bronze technology diffused not only westward to Europe but also eastward. Bronzes with high levels of tin have been found at Mohenjo-Daro in the Indus Valley from 2500 B.C. and in China by about 2000 B.C. Smelting may have been discovered independently in China, as all the necessary raw materials were readily available and kiln temperatures of 1200°C were in use by potters. Moreover, the western and American techniques of hammering or forging were entirely absent. If bronze technology had come by diffusion, these techniques should have been known.

Figure 7.8. Bronze plaque (5 inches high) from fifth or fourth century B.C. *Greece, showing a satyr. Virginia Museum of Fine Arts, Richmond, VA. The Adolph D. and Wilkins C. Williams Fund. © 1988 by Virginia Museum of Fine Arts.*

Chinese technology seemed to begin with casting. The Chinese tradition led to a very high artistic level by the eleventh-century B.C. Shang Dynasty (Figure 7.9). As with European bronze, later Chinese bronze also took on added levels of lead, by comparison, for example, of Shang, Zhou, and Han Dynasty bronzes. Discoveries of supposedly very early bronzes in Ban Chiang and elsewhere in Thailand proved to have been compromised by mixed stratigraphy. Nonetheless, a bronze industry had been established in Thailand and Vietnam by about 2000 B.C.

A possibly independent tradition arose in West Africa, near modern Nigeria, around the ninth century A.D. A set of bronzes from the site of Igbo-Ukwu

Figure 7.9. Chinese bronze Yu (ceremonial wine vessel, 14 inches in height) from the Shang Dynasty (Anyard period, thirteenth–eleventh centuries B.C.). Anna Mitchell Richards Fund. Courtesy of the Museum of Fine Arts, Boston.

were excavated during the 1960s and analyzed by Thurstan Shaw in the 1970s and by Paul Craddock in the 1980s. Craddock commented that "despite the long tradition of metalworking manifest in them, they are without known precedent or parallel." The unique style suggests the absence of European influence. Casting was entirely by the lost wax method. Analysis of the alloy found high levels of tin (6–12 percent) and lead (1–10 percent), little or no zinc (<0.5 percent), and significant levels of silver (0.1–0.9 percent). The consistent profile of trace elements, particularly of silver, suggested that the copper ore had come from a single location rather than being traded in from numerous sources. Craddock's analysis supported an independent West African tradition that may have dated to as early as 2000 B.C., when furnaces found in Niger were able to melt out nodules of native copper present in copper ore. The West African tradition, however, had entirely died out by the arrival of Europeans. Craddock suggests that Arab trade from the eleventh century A.D. may have rendered the local tradition obsolete.

Bronze in the Americas

Although it is unclear whether copper smelting was discovered independently in several Old World regions, it is certainly the case that all metallurgical discoveries in the Americas prior to European contact were independent. Indeed, metallurgical evolution followed entirely different pathways in the New World. Thus there is evidence for up to five independent discoveries of smelting: Southwest Asia, southeastern Europe, China, West Africa, and South America. It has already been noted that an abundance of native copper led to cold-worked and annealed copper artifacts during the fourth millennium B.C. in North America but never resulted in melted or cast copper. Hammered gold foil artifacts have been found from as early as 1500 B.C. from the Andean highlands of Peru. There is clear evidence for copper mining by 1000 B.C., also in the Peruvian highlands. The gold and copper traditions were brought to great heights by the Chavín Culture in South America. One of the most remarkable early artifacts is a small bead found at the central Peruvian site of Malpaso made of about 45 percent silver and 41 percent copper, an alloy that could have been formed only by mixing metals in the molten state. Silver itself melts at 960.5°C. Thus temperatures of about 1000°C were achievable after around 1000 B.C., and the advantages of the molten state were understood for the first time in the Americas.

It is likely that the American production of copper metal by smelting began soon after this discovery. The Chavín focus, however, was on gold and its alloys. The Mochica Culture, on the north coast of Peru, was smelting copper ores on a larger scale beginning as early as 200 B.C., and they were able to produce pure silver and lead as well. Copper was hammered and cast, both in open molds and by the lost wax method. Mochica alloys included gold-silver, copper-silver, and copper-gold. These alloys of copper did not make the object harder or stronger but rather changed its color. Thus much of the metallurgy of South America was directed toward aesthetic rather than practical ends.

The Mochica also discovered arsenical copper, although its use was only occasional. Their successors, the Chimú, made wide use of the alloy (2–6 percent As) for implements such as needles, chisels, and agricultural tools, during the Late Intermediate Period, after A.D. 1000. These materials were made possible by rich sulfarsenide ores of copper throughout the Peruvian highlands, such as tennantite, $Cu_{12}As_4S_{13}$. On the other hand, the cultures in southern South America, including Bolivia, Argentina, and Chile, developed an entirely distinct metallurgy based on the alloy of copper and tin (tin bronze). The richness of the Bolivian tin mines is still being exploited today. The southern Andes artisans both cast and hammered bronze. Thus two distinct but parallel regional technologies developed after A.D. 1000, with arsenical copper in the northern Andes and tin bronze in the southern Andes. These traditions were brought together after 1460 by the all-conquering Incas, who preferred tin bronze.

The Liberty Bell

Although this book is concerned primarily with archaeological materials, the American Liberty Bell is probably the most famous bronze object in the United States, and its history is basically a chemical story. The analysis of archaeological and of historical objects employs a common methodology. Moreover, the story of the Liberty Bell provides considerable insight into the subtleties of the metallic proportions. It begins in 1751, when the Pennsylvania Assembly commissioned the Whitechapel Foundry of London to cast a bell for the new State House in Philadelphia. It was then known that the optimum bronze for a bell contained about 77 percent copper and 23 percent tin. Lower levels of tin produced too soft a metal and higher levels too brittle a metal for regular use as a bell. When it was installed in Philadelphia, it cracked on its first ring. In standard litigious fashion, the Philadelphians said the metal was too brittle, while Whitechapel said that the bell-ringer was an amateur. A local brass founder named John Stow, who had no experience with bells or large objects, was asked to recast it. He must have thought that the tin level was too high, as he purchased copper metal to add to the melt. The second bell had such a disagreeable sound that Stow was asked to cast it yet again. It is likely that he increased the tin level this time. The third version is the bell that was used to proclaim the signing of the Declaration of Independence on July 8, 1776. It was used continuously until 1835, when it cracked while tolling the death of Supreme Court Justice John Marshall.

Complete analysis of the metal was carried out during the 1970s by the Winterthur Museum in Delaware. The average copper level was found to be only 67.1 percent and the tin level 26.9 percent, clearly deviating from the optimal composition and indicating a brittle alloy. Moreover, their thirteen measurements had a tin range of 25.20 to 30.16 percent, a considerable variability that probably arose because Stow had no equipment for large melts. His experience with horse bells and rivets involved only small vessels, so that the molten bronze for the bell would have been poured into the mold in stages with variable composition. The alloy also had 1.4 percent zinc and 3.25 percent lead, which may have come from scrap metal used by Stow to adjust the Cu (copper) and Sn (tin) levels in the two recastings. The unfortunate history of the Liberty Bell thus resulted from an inappropriately high level of tin.

Corrosion of Bronze

The archaeologist must deal with objects that may have degraded over time. The strength of the pure metal is compromised when it returns to its oxidized forms through the process of *corrosion*. When bronze objects are underwater or in the soil, corrosion is relatively slow, as the process requires oxygen. Exposure to air can lead to formation of a patina of tenorite (CuO), cuprite (Cu_2O), or the familiar green malachite ($Cu_2CO_3(OH)_2$), which also requires the presence of atmospheric carbon dioxide. The most corrosive activity occurs when chloride ion, as from salt ($NaCl$), is present along with oxygen. Initially copper metal is converted to its charged copper(I) form (Cu_2Cl_2), which reacts with oxygen to form the more stable copper(II) form ($CuCl_2$). Copper(II) chloride can react with new copper metal, according to Equation 7.3, to produce more copper(I) chloride and thereby start a chain reaction.

$$CuCl_2 + Cu \longrightarrow Cu_2Cl_2 \qquad (7.3)$$

Powdery green spots on the surface of bronze objects have been called bronze disease, and they can continue to grow as long as water and air are present. The condition may be slowed by proper control of humidity, but complete cures are highly invasive. The corroded part may be removed by electrolysis or physical excavation, and the scar packed with silver oxide (any remaining chloride reacts with silver to form the stable silver chloride). An entire object suffering from the condition may be soaked in sodium sesquicarbonate until all chloride ion has been removed.

Sources of Tin

The critical component of bronze, of course, is the added element tin. Ample local supplies in Bolivia and China encouraged the respective early development of bronze in South America and East Asia. The principal tin ore almost everywhere was cassiterite (SnO_2), an easily smeltable material by reductive techniques. The Greeks and later the Romans had reliable supplies from Cornwall, Brittany, and Iberia. These mines had been developed from as early as 1000 to 800 B.C. The question of the tin source becomes more difficult during the earlier flowering of Aegean and Egyptian civilizations. For many decades it was thought that their tin came from Cornwall via Phoenician traders, but the Cornish mines had not yet been opened in this period from 2000 to 1000 B.C. R. D. Penhallurick suggested that the source was the rich tin supplies from central Europe, in the Erzgebirge mountain chain that forms the present border between Germany and the Czech Republic. Trade from the north down to the Aegean is documented by the presence of amber in numerous second-millennium B.C. Mediterranean sites. By his infrared method, Curt Beck found that 230 out of 264 pieces of amber from Mycenaean and Helladic sources (1550–1050 B.C.) were of Baltic origin, 32 were indeterminate, and only 2 were clearly non-Baltic. The association of amber and tin is exemplified by the extraordinary necklace shown in Plate 15 that is made up of beads of pure tin, amber, and glazed quartz. This object brings together metal, stone, and glaze/glass technologies in a dramatic unity. It was found in a peat bog in Holland in 1881 and dates to about 1400 B.C. If amber could come from the north, the argument goes, so could tin. Although not proved, the central Europe connection currently is the best hypothesis.

The Erzgebirge sources, however, were not available to the Southwest Asians at the dawn of the use of bronze in the fifth and fourth millennia B.C. The very slow development of bronze technology over these several thousands of years may have been related to the poor availability of tin. Objects entirely of tin have been found in Egypt (a bottle from about 1500 B.C.) and on the Aegean island of Lesbos (a bracelet from about 3000 B.C.). Prior to these dates, tin is known only as a component of bronze. Considerable light has been shed on this problem by the discovery during the 1980s of tin mines in Anatolia. Ashlihan Yener and her Turkish co-workers have accumulated evidence that the site of Kestel in Anatolia was used to mine tin from at least 2870 B.C. (Figure 7.10). (Kestel was located north of the coastal city of Tarsus and not far from the famous later copper mines of Cyprus, a name that probably comes from the same root as the word *copper*). The ore grade was very low, but Yener demonstrated experimentally that tin could have been smelted from it. She also found crucibles and slag that appear to have been involved in

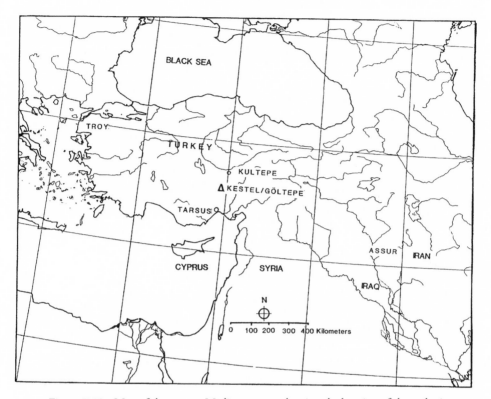

Figure 7.10. Map of the eastern Mediterranean, showing the location of the early tin mine at Kestel.

the processing of tin ore. She found tin ore as well in Bolkardağ in the Taurus Mountains. The hypothesis that has evolved, not without controversy, is that numerous tin sources in Southwest Asia contained small pockets of low-grade ore and could be exploited for the early production of bronze. Their geographical proximity to the first smelting of copper metal makes them a likely source for the earliest use of tin as an intentional alloying component. Their rarity and poor grade slowed the development of bronze metallurgy, which did not flower until well after 3000 B.C., possibly when the central European sources of tin became available.

Lead in Antiquity

The most common metal after tin to be added to copper was lead. It has already been noted that the early use of lead metal might serve as a possible indication for the invention of copper smelting. These finds date to the sixth millennium B.C. and demonstrate that lead smelting was one of the earliest metallurgical processes. Noël Gale and Zofia Stos-Gale have suggested that lead smelting may even have preceded copper smelting. For one reason, lead may be smelted at 800°C in a very simple cooking fire. The process may have been carried out very casually, in a small open fire prepared for the purpose and leaving very little archaeological residue.

The most common lead ore is its sulfide, galena (PbS). In an oxidizing atmosphere, the sulfide is roasted to the oxide, according to Equation 7.4.

$$2PbS + 3O_2 \longrightarrow 2PbO + 2SO_2 \qquad (7.4)$$

The lead oxide (PbO, also called *litharge*) then is reduced by the residual unroasted ore (Equation 7.5)

$$2PbO + PbS \longrightarrow 3Pb + SO_2 \qquad (7.5)$$

or by carbon introduced for the purpose (Equation 7.6).

$$PbO + C \longrightarrow Pb + CO \qquad (7.6)$$

The melting point of lead is 327°C, but smelting occurs at 800°C, so that lead is formed in its liquid state.

During the Early and Middle Bronze Age, lead found few uses, either as a pure metal or as an alloying component of bronze. J. O. Nriagu estimated that the total worldwide consumption of lead up to 2100 B.C. had been only a million tons. The principal uses were for small items such as spindle whorls, sinkers for fishnets, rings, beads, bracelets, small statues, and weights, and as a material for repairing pottery. Almost all the large number of lead weights found in Bronze Age sites on the Aegean islands of Kea and Thera were in multiples of 61 grams, indicating that a uniform system of units existed at this time, possibly over the entire Aegean/Minoan world.

As lead was added to tin bronze during the later stages of the Bronze Age (2100–1200 B.C.), its use increased dramatically, with most of the lead produced in the Aegean Islands and Southwest Asia. Iberian sources were added during the Iron Age. The Greeks were the first to include lead in the bronze alloy used for statues.

Production levels during the Roman Empire (through A.D. 500) were so high, peaking at about 80,000 metric tons per year, that lead has been referred to as the Roman metal. In addition to its use in bronzes, the pure metal and its alloys with tin found uses on an industrial scale: in coinage, roofing, vat linings, pipes, coffins, kitchenware, and the linings of the hulls and keels of ships. The Latin word for lead is *plumbus*, from which comes not only the chemical abbreviation (Pb) but also the English words *plumber* and *plumb*.

The alloy of tin with 5–25 percent lead is called *pewter* and has been widely used for tableware. A lead-rich tin alloy (about 60 percent lead) has been called *ganza* by Europeans in Southeast Asia, where it was used for coinage. A similar tin alloy (50–70 percent lead) has very low melting properties and is used widely as solder for binding two metals together.

The upper plot in Figure 7.11 shows an estimate of lead production on a worldwide basis over the last five thousand years. Most lead probably was produced as a by-product of silver refining. After the peak during the Roman Empire, production dropped almost 90 percent and attained Roman levels only during the Industrial Revolution. A group headed by Claude F. Boutron tracked the lead emissions of this industrial activity by measuring lead levels in ice core samples from central Greenland. They established preindustrial levels from a 7,760-year-old core. The lower plot in Figure 7.11 tracks lead levels in the ice, presumably reflecting levels in the atmosphere. The level quadrupled during Roman times, then settled back almost to preindustrial levels during

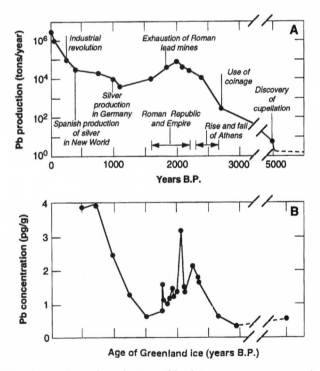

Figure 7.11. Top, *estimated production of lead in tons per year over the last five thousand years (B.P. = before present)*. Bottom, *concentration of lead in central Greenland ice, with one point from 7,760 years ago, and the remaining points between 2,960 and 470 years ago. Reprinted with permission.* © *1994 by the American Association for the Advancement of Science.*

the Middle Ages. Levels then increased rapidly after A.D. 1000. Lead was produced primarily in open air furnaces, with emissions passing into the atmosphere and spreading even as far as the Greenland Arctic. Thus human pollution of the atmosphere, at least in the Northern Hemisphere, had begun three thousand years ago.

The wide use of lead in the ancient, particularly Roman world may have led to widespread lead poisoning, or plumbism. There is no question that the element is highly toxic in many chemical forms. Occupational exposure by miners, smelters, and manufacturers may have caused acute or chronic plumbism. Diagnosis is difficult because its symptoms resemble those of a viral infection: headache, nausea, abdominal pains, and listlessness. The Romans in particular were highly susceptible because of nonoccupational exposure from lead pipes used to carry water and lead vessels used for food and wine. Lead found numerous such uses because it is extremely stable to corrosion. Whether it truly contributed to the decline of the Roman empire is possible but uncertain. The Roman aristocracy certainly experienced very low fertility rates (by the second century A.D., out of all the famous Republican patrician lines, only the Cornelii family had survived). The only direct evidence has been measurement of lead levels in skeletons. In preindustrial societies, the lead burden is less than 5 parts per million (ppm). Studies by H. A. Waldron and A. Mackie found levels in Romano-British skeletons ranging from 64 ppm at York to over 300 ppm in Cirencester. These high levels are suggestive, but not definitive, that lead poisoning was a widespread Roman malady.

Sources of Lead

An effective procedure for identifying the provenance of stone artifacts is analysis of the proportions of trace elements. Except with native copper, this approach is not feasible for the study of metals, because the process of smelting can significantly alter the natural profile of trace elements. Moreover, remelting and mixing previously used metals during casting could lead to mixed profiles. A quirk of the geochemistry of lead, however, permits the alternative method of isotope analysis to be used to determine sources of lead (described earlier for the analysis of marble, majolica glaze, and glass). The radioactive decay of uranium and thorium contributes significantly to the natural levels of the lead isotopes with atomic weights 206, 207, and 208. Only lead-204 is nonradiogenic in origin. Because ores vary in their proportions of uranium and thorium, the proportions of the four isotopes vary geographically.

For twenty-five years, several groups have been characterizing mines by their lead isotope ratios and relating metal artifacts to their mining sources by the isotope content. Contributors include Noël Gale, Zofia Stos-Gale, Robert Brill, and Ernst Pernicka. Isotope levels are usually expressed as ratios with respect to lead-206, so that three isotopic numbers are available for each mining site (208/206, 207/206, and 204/206). Three-dimensional plots have been used, but more commonly the data are exhibited in two two-dimensional plots such as those in Figure 7.12, in which many of the major Mediterranean lead-containing copper sources are illustrated. These sources may be found on the map in Figure 7.13. There is clearly a lot of overlap of sources, but the principal ones (Cyprus, the Aegean island Kythnos, and the mainland Greek site Laurion or Lavrion) are well separated on one or the other plot. Alternatively, the ratios may be combined to create two composite factors (canonical variables or principal components) for a plot such as that in Figure 7.14, exhibiting excellent separation.

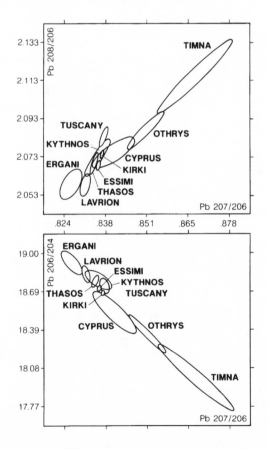

Figure 7.12. Separation of lead-containing copper ore sources from the Mediterranean according to plots of four lead isotope ratios. Courtesy of the British Academy.

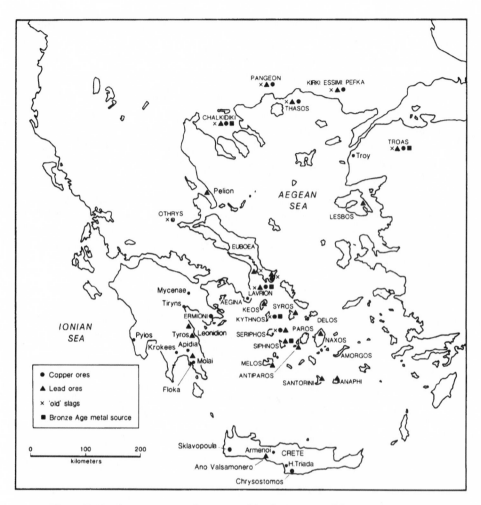

Figure 7.13. Aegean sources of copper and lead. Courtesy of the British Academy.

From plots of these types, the lead sources of the eastern Mediterranean have been characterized thoroughly. Gale and Stos-Gale found that Early Bronze Age artifacts contain lead mined from the Aegean islands of Kythnos and Siphnos, as well as from the mainland site of Laurion. Other islands, however, also had ore deposits. The Laurion mines were the source for lead found at Knossos on Crete from Middle Minoan to Mycenaean times. Three caldrons found by Heinrich Schliemann at Mycenae not surprisingly contain lead from the Laurion mines. Even some Egyptian artifacts from Amarna and Abydos (Eighteenth Dynasty) proved to derive from Laurion. Thus by Late Bronze Age times, the Laurion mines were the overwhelmingly largest source

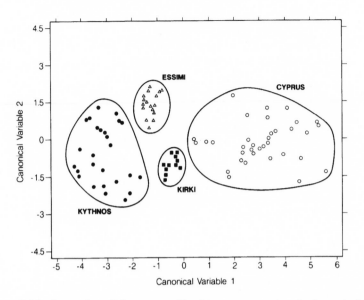

Figure 7.14. Separation of the sources of lead in copper ore deposits from the Mediterranean according to plots of lead isotope ratios represented by two principal components. Courtesy of the British Academy.

of lead. Although the mines on Cyprus were becoming increasingly important, they were not supplying these Greek and Aegean sites with a large proportion of their lead.

Isotope techniques were applied to the bronzes from Igbo-Ukwu and the nearby but later sites of Ife and Benin in West Africa (Figure 7.15). Craddock and his co-workers from Nigeria found that the lead isotope ratios of Igbo-Ukwu bronzes corresponded to those found for galena ore from mining sites in the Benue Rift deposits of Nigeria itself. The attribution of the lead to a local source strongly supports their theory of the independent, indigenous development of metallurgy in West Africa.

Zinc in Antiquity

The last major elemental addition to copper was zinc. The alloy of copper and zinc is called *brass*, and the earliest object with such a composition may be a bracelet containing 11 percent zinc and no tin from the Assyrian site of Cavustepe in eastern Turkey, dating to about the eighth century B.C. This object is considered to be a natural brass, prepared from a copper ore with an accidental zinc impurity. It is not certain whether intentional brass arose first in the Greek world, Southwest Asia, Egypt, or India, but it appeared during

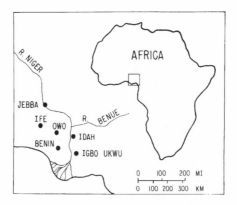

Figure 7.15. Sites of major Nigerian metalworking traditions. Reprinted with permission. Copyright 1978 by the American Chemical Society.

the last third of the first millennium B.C. in the Hellenistic world and in considerable quantity in the Indian subcontinent. Brass objects from the fourth century B.C. have been found at Taxila in Pakistan, and brass coinage first appeared in Bithynia and Phrygia in Southwest Asia at the beginning of the first century B.C. Zinc was present in Chinese bronzes as early as Han times, after 200 B.C. Brass Roman coinage began about 45 B.C., and the zinc content was later supplemented with tin to form the alloy called *gunmetal.* There may have been an increased scarcity of tin, which certainly was the case after the loss of the western Roman Empire to the barbarian invaders after the fourth century A.D. Without the tin sources of Hispania, Britannia, and Pannonia, eastern Europe and Southwest Asia turned from bronze to brass, which continued in popularity during the medieval period in Europe.

The color of brass containing 10–18 percent zinc is almost golden, and the presence of zinc conveys the same properties of strength and hardness as do arsenic and tin. Its glitter has been particularly popular in India, where brass has been in use continuously for two thousand years, in roofing for temples, furniture, and cooking and storage vessels.

There were two distinct metallurgical processes in antiquity for manufacturing brass, both largely defined by the volatility of zinc. The pure metal melts at 420°C and boils at 917°C, well below the smelting temperatures for copper. The earliest method probably was the *cementation process,* in which particles of copper are heated to about 1000°C in the presence of zinc oxide (either the ore or the product from roasting zinc sulfide) and charcoal in a closed vessel. Zinc vapor is formed by reduction (Equation 7.7) and is dissolved into the copper, so long as the atmosphere is kept reducing by the presence of carbon monoxide. Modern experiments have demonstrated that this process can produce brass with no more than 28 percent zinc, a barrier that remained in the European world almost until modern times. The presence of tin or lead in the ore reduces the ability of copper to dissolve zinc, so that the maximum percentage of zinc then is below 28 percent. When the raw material is zinc carbonate ($ZnCO_3$, smithsonite or calamine, hence the *calamine process*), the chemistry is quite similar, as the carbonate loses carbon dioxide to form zinc oxide.

$$ZnO + C \longrightarrow Zn + CO \qquad\qquad (7.7)$$

Brass also can be made by the direct fusion of metallic copper and zinc, a process predicated on the availability of pure zinc metal. This method permits zinc levels of much more than 28 percent, although when above 46 percent the alloy becomes brittle. It may have been used in the Indian subcontinent from the earliest times, as a brass vase from the Bhir mound of fourth-century B.C. Taxila in Pakistan was found to contain 34.3 percent zinc. Recent excavations at Zawar in western India (Figure 7.16) have found strong evidence for the production of zinc metal for more than two thousand years. The mines at Zawar were already 100 meters deep two thousand years ago, and the amount of debris suggested that over a million tons of zinc were produced.

Excavation of the actual furnaces at Zawar has revealed the details of the smelting process, as described by Paul Craddock. The ore was sphalerite (ZnS), which first was roasted to zinc oxide. The zinc oxide was packed with a fuel (a reducing agent such as a resin) and possibly with common salt as a fluxing agent to lower the melting points of silica and alumina. The charge was placed in a retort and inverted in the furnace (Figure 7.17). During smelting, the zinc vapor moved downward under a carbon monoxide atmosphere (to prevent oxidation back to zinc oxide) to a cooler region, where it condensed into receiver flasks.

An analogous process to generate metallic zinc was not available in Europe until after the Industrial Revolution. In postmedieval Europe, improvements in the cementation process were able to increase the percentage of zinc up to 33 percent, which became a common composition during the eighteenth century. The standard composition for nineteenth- and twentieth-century western brass has been about 34 percent zinc.

Knowledge of acceptable proportions can be used to test authenticity of allegedly ancient materials. In 1579 Sir Francis Drake may have landed on the coast of California during his voyage around the world. In 1936 a plate of brass was found in Marin County, in the San Francisco Bay Area, that resembled a description of a plate supposedly left by Drake. Its inscription included "BEE IT KNOWN VNTO ALL MEN BY THESE PRESENTS IVNE 17 1579 . . . I TAKE POSSESSION OF THIS KINGDOME WHOSE KING AND PEOPLE FREELY RESIGNE THEIR RIGHT AND TITLE IN THE WHOLE LAND . . . NOW NAMED BY ME . . . AS NOVA ALBION . . . FRANCIS DRAKE." Although early analyses claimed to substantiate the authenticity of the plate, the work by H. V. Michael and

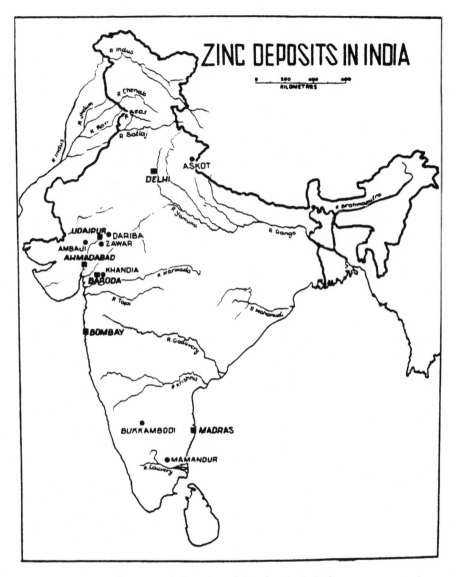

Figure 7.16. Location of zinc deposits in India.

F. Asaro found a composition of 64.6 percent Cu (copper), 35.0 percent Zn (zinc), 0.10 percent Pb (lead), 0.27 percent Fe (iron), 0.12 percent Cd (cadmium), and 0.006 percent Sn (tin). The levels of the two major elements are typical for the late nineteenth and early twentieth centuries and were essentially unknown in the sixteenth century, while the level of zinc exceeds both the traditional 28 percent level and the then-absolute 33 percent maximum. Moreover, the composition was entirely uniform and lacking in trace elements

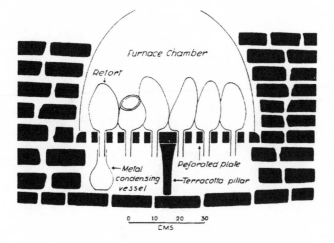

Figure 7.17. Diagram of a furnace for smelting of zinc at Zawar, India.

of more than negligible levels. Typical brass objects of the fifteenth and sixteenth centuries contained 24–30 percent Zn (zinc), 1.2–3.0 percent Sn (tin), and 2–7 percent Pb (lead). It seems that sometime not long before 1937, an enterprising metalsmith prepared and deposited the object for the public to find.

In West Africa an extended flowering of brass production followed the Igbo-Ukwu bronze production. The city of Ife (Figure 7.15) created an artistic tradition of casting heads, primarily in brass, whose naturalistic and refined appearance (Figure 7.18) is unique. Because the copper did not contain the consistent silver impurity of Igbo-Ukwu bronzes and because the alloy had largely shifted from bronze to brass, Craddock concluded that the metals had come from Arab trade rather than from the indigenous Nigerian sources. After about 1500, similar materials were produced in neighboring Benin, although the heads were increasingly less natural. An oral tradition named the Ife smith as Igueghae and stated that Ife provided the technology at about the time that the Ife royal family took over Benin. The heads may be portraits of these chiefs from life. All the heads were cast by the lost wax method, in the tradition of Igbo-Ukwu. Although Europeans originally thought the "Benin bronzes," as they were called, were of nineteenth-century origin, inspired by European contact, they in fact were an independent artistic creation based on the ancient metal technologies of West Africa.

The Iron Age

Iron began replacing copper as the principal alloying metal in the Mediterranean world during the second millennium B.C. It is usually suggested that

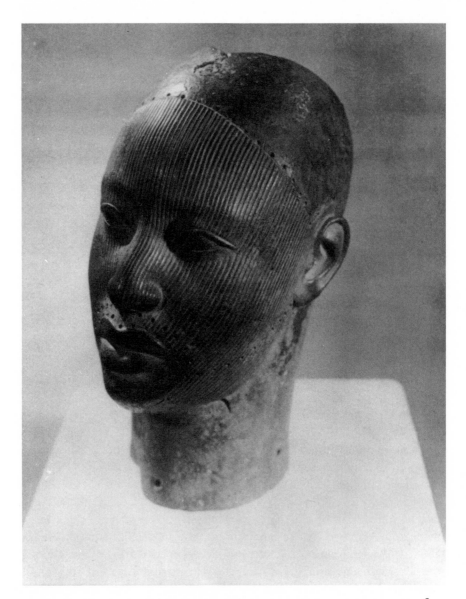

Figure 7.18. Brass head from Ife, Nigeria (Ife Museum Cat. No. 20), height 12 $\frac{5}{8}$ inches.

supplies of copper or tin were being mined out, so that new types of raw materials were needed. Alternatively, the properties of iron alloys may have been recognized to be superior, or the wide availability of iron ores may have made the new metals economically preferable. The necessary smelting technology had to be in place before the Bronze Age could move into the Iron Age. These changes did not occur until about 1200 B.C., probably in the Anatolian peninsula populated by the Hittite people. At this time iron began to be alloyed with carbon to produce the much stronger material called *steel.*

As with copper, the first source of iron was the native metal, most widely available from meteoritic sources. This iron contains at least 4 percent nickel and hence is readily identifiable by chemical analysis. Jane Waldbaum in 1980 stated that only fourteen iron objects were known from before 3000 B.C., all from Iran, Mesopotamia, or Egypt, and most of those analyzed were clearly meteoritic. From the next fifteen hundred years, corresponding to the Early and Middle Bronze Age, iron objects are found sporadically but increasingly in these areas as well as in Anatolia. The Early Bronze Age site of Alaca Hüyük in particular has produced numerous such objects, including ceremonial weapons and ornaments. Chemical analysis indicates that about as many were smelted as came from meteoritic sources.

The increased production of iron for utilitarian purposes, such as weapons and tools, after 1500 B.C. has traditionally been associated with the Anatolian Hittites. The approximate year of 1200 B.C. has been assigned to the beginning of the Early Iron Age, but few peoples at this point possessed a predominantly iron metallurgy. The decline of the Hittite culture may have encouraged the dissemination of iron technology. After 1000 B.C. it spread around the Mediterranean, to Hallstatt Europe by about 800 B.C. and to Britain by about 500 B.C. The Nok culture in Nigeria was able to smelt iron by 400 B.C., and it is possible that it was an independent invention. Iron technology never appeared in pre-Columbian America. It developed in China in an entirely different fashion and possibly was an independent invention there as well. Whereas early Western societies forged iron below its melting point, the Chinese, before 500 B.C., developed a method of melting and casting iron. Chinese iron artifacts date back to the Shang period. By the Han Dynasty cast iron objects were common, including coins, bells, pans, knives, and chariot wheel axles.

Processing Iron

Smelted iron probably was discovered through the smelting of copper. Iron ores were used as fluxes in copper smelting, under which conditions metallic iron could have been produced and recognized as a new type of metal. Then the artisan had to deduce which raw material was responsible for the new metal. The melting point of iron, at 1540°C, is well out of the range available to Bronze Age fires. The reduction of solid iron oxides, however, occurs at about 800°C, so that smelting is carried out in the solid state. The actual smelting temperature is about 1200°C in order to form a liquid slag (silicates and aluminates) that in part could separate out. The process is very inefficient, and Iron Age slag generally contained as much as 50 percent iron and in later millennia was used as a raw material for iron. Charcoal present as fuel provided the reducing agent, represented as carbon monoxide in Equation 7.8 for the smelting of the ore hematite.

$$Fe_2O_3 + 3CO \longrightarrow 2Fe + 3CO_2 \qquad (7.8)$$

The product of this process was a spongy mass resembling a flower and hence was called a *bloom,* composed of elemental iron, slag, and unreacted charcoal. Pieces containing concentrated iron were broken out of these lumps and then formed into a new mass on the smith's fire. This mass was forged by heating and hot-hammering, thereby forcing out the slag and charcoal. The eventual product of several cycles of such forging was *wrought iron,* which contains less than 0.08 percent of carbon. Wrought iron may have served as a substitute or replacement for bronze, but its properties were not necessarily superior.

The discovery of steel before 1200 B.C. involved the realization that heating iron in glowing charcoal yielded a metal that was stronger than wrought iron and retained a sharper cutting edge. The identification of carbon as the key alloying element, however, did not occur until the nineteenth century. Steel normally contains 0.2–0.7 percent (and always under 2 percent) of carbon. It was the advantages of steel that really set the Iron Age into motion.

Further hardening later was achieved by heating the object to above 750°C in a reducing atmosphere (to avoid burning out the carbon) and quenching the object in cold water to freeze in the internal metallic structure. Some annealing (or tempering) of the quenched metal by mild reheating was needed to minimize brittleness. Each of these steps, from iron ore to bloom to wrought iron to raw steel to quenched steel to tempered steel, may be followed by structural changes revealed under the microscope.

Like the early copper furnaces, the first iron furnaces were depressions in the ground or in a rock, called bowl furnaces. Without an outlet, the liquid slag just descended to the bottom of the bowl, with the bloom on top of it. By Roman times, slag tapping was carried out by means of a runoff passage at the bottom that led to a side hole. The next stage of development involved the introduction of shafts over the bowl to improve the draft and help maintain reducing conditions (Figure 7.19).

Examination of the earliest iron objects from China indicates that Chinese metallurgy involved melting and casting iron, presumably by means of a primitive *blast furnace*. A feasible melting point (1150–1200°C) was achieved by the presence of significant amounts of impurities in the iron: 2–5 percent carbon, 1–3 percent silicon, as well as phosphorus and other elements. Actual melting of pure iron at 1540°C was not possible. It was primarily carbon and phosphorus that lowered the melting point, and silicon served to enhance its castability. The alloy generally referred to as *cast iron* consequently has a high level of impurities, typically 4 percent carbon, much higher than in steel. The Chinese blast furnace utilized efficient bellows, which by A.D. 50 were powered by a water wheel. The European blast furnace may have been developed as early as A.D. 1200 in Sweden, but it came into common use during the fifteenth century, with a concertina bellows to supply air (the blast) and a water wheel to provide the mechanical force to drive the bellows. The efficient introduction of air enabled these furnaces to achieve higher temperatures and more reducing conditions than bloomery furnaces, and hence to achieve melting of the impure iron. Cast iron is more brittle than wrought iron or steel because of the higher levels of impurities. The carbon could be removed almost entirely (decarburization or fining) through heating in an oxidizing blast. This purified iron could be converted to steel by the controlled reintroduction of carbon (carburization).

In Africa the Early Iron Age began by about 500 B.C. Although the technology may have arrived from Europe or Asia by diffusion, alternatively it may

Figure 7.19. Greek vase from the sixth century B.C., depicting a smith and a bloom furnace.

have been invented independently and certainly took a unique turn. Fuel from plants used in Africa left large amounts of phosphorus in the smelted product in addition to carbon. The high levels of these impurities led to a brittle metal that was strengthened by decarburization. This fining procedure removed carbon but left the phosphorus impurities. Removal of carbon normally would produce wrought iron, but phosphorus can fulfill much the same role as carbon. The phosphorus steel thus produced in Africa was both malleable and strong and has a structure found nowhere else. The various types of iron alloys are reviewed below:

Type of material	% carbon	% phosphorus
Wrought iron	<0.1	<0.1
Steel	0.2–2	<0.1
Cast iron	2–5	<0.1
Phosphorus steel	<0.2	0.2–4

Iron and steel were put into use primarily for applications requiring strength and durability. Nowhere were the changes more effective than in warfare, as the steel sword became the major weapon. From the eleventh to the seventh centuries B.C., people in the Luristan area of Iran were manufacturing magnificent steel swords with 0.2–0.5 percent carbon, which, however, had not been quench hardened. By the third century A.D., a new technique called *pattern welding* was developed, possibly in the Roman Rhineland. Rods and wires of iron or steel were twisted together and forged. The product was folded and forged again, and the process repeated many times (Figure 7.20). Patterns were caused by weld lines, slag entrapment, or iron of varying composition and could be carried to an art form, as exemplified by the sword found in the Saxon Sutton Hoo burial and by numerous Merovingian swords. Variations on the technique produced the samurai sword in Japan and the curved kris blade in Southeast Asia (Figure 7.21, p. 202).

An alternative technology was developed in India about A.D. 1000. Wrought iron was placed with some organic materials in small crucibles that were sealed and fired to about 1200°C, until the iron absorbed enough carbon to melt out as steel, usually 1–2 percent carbon. The product has been called *crucible steel* or *wootz*. It was the primary raw material for Arab artisans in Persia and Syria. Swords produced from wootz by a process similar to pattern welding entered the European market from Damascus and became known as *damascene*. The metal exhibited characteristic wavy surface markings called damask caused by variations in the carbon contents. The damascene product was produced by hammering layers together without the twisting associated with the European blade.

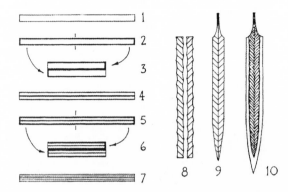

Figure 7.20. Stages in construction of a pattern-welded sword. An iron rod (1) is heated (2), folded (3), and hammered out (4). The process is repeated many times (5, 6, and 7). Two such rods are twisted (8) and welded (9) to form the blade, which is provided with an edge, polished, and etched to bring out the pattern (10).

Gold in the Old World

Copper and iron provided the dominant alloying metals for the ancient world, with both utilitarian and ornamental applications. Gold, however, may have been the earliest metal used by humans, because of its availability as the native metal, its extraordinary beauty, and its resistance to corrosion. Within the last two hundred years the fever of its attraction has drawn multitudes of people to discoveries in Australia, South Africa, California, and Alaska. The native metal is usually found in secondary alluvial deposits (those caused by running water), into which it had washed from primary deposits and from which it can be recovered by placer mining methods. Sometimes it is mined from primary deposits (the mother lodes), found usually as veins in quartz. Some smelting procedures may then be required. The ancients developed effective methods of separation. As described by both Vitruvius and the elder Pliny, gold could be recovered by reacting the ore with elemental mercury to form gold amalgam, which could be removed and heated to drive off the volatile mercury and leave a residue of pure gold.

The early classical civilizations benefited from convenient and rich sources of gold, including Egypt, the Aegean Islands, Macedonia, Nubia, Spain, Gaul, and India. Native gold contains almost no other elements except silver, which can be present from 5–50 percent. Above about 20 percent, the alloy is referred to as *electrum*, a term used in the ancient world also to refer to the synthetic alloy of the two elements. The process of removing undesired silver from gold is called *parting*. From late medieval times, parting involved treatment with nitric acid to remove silver preferentially, but today chlorine gas is bubbled through molten impure gold. In antiquity, however, the process was carried out in the solid state by a cementation process that dates back to before the second century B.C. The impure metal was hammered into thin sheets, cut into small pieces, and placed in a crucible along with salt (NaCl), silica, or alumina as a flux, and an acid such as urine. When the mixture was heated to just below the melting point of the metal, chloride reacted with the

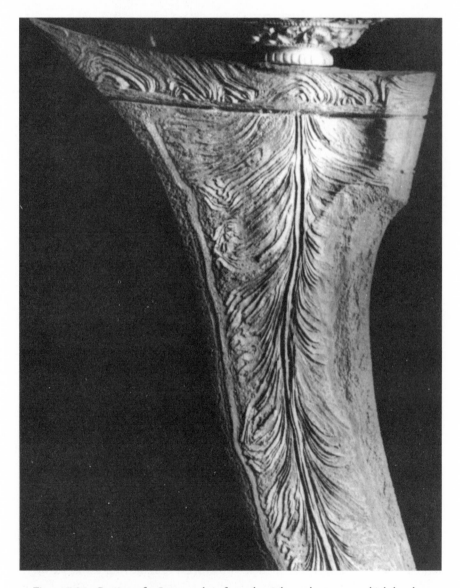

Figure 7.21. Portion of a Javanese kris from the eighteenth century, which has been etched to bring out the patterns. Courtesy of the Seattle Art Museum.

acidic flux to form hydrochloric acid or chlorine gas. These materials reacted with silver to form silver chloride, which is volatile at the reaction temperature and is absorbed into the flux or the crucible walls. Pure gold was left, and the silver often could be extracted from the vessel.

Although objects could be made entirely of gold, the value and scarcity of the metal led ancient artisans to develop numerous methods to give an object the appearance of gold, a process generally referred to as *gilding*. One of the earliest gilding techniques used in the Old World was *granulation,* whereby many very small beads of gold were attached to an object until the surface was covered. The Sumerians were practicing this technique by 2000 B.C. A related method used fine gold threads and was called *filigree*. The object probably was coated with glue, often containing a copper compound as solder, and the granules or threads were set onto the sticky surface. The object was heated to about 900°C. Copper oxides in the solder, in essence, were smelted by carbon in the fuel or glue to produce metallic copper, which formed a bond between the object and the attachments. Both of these techniques went into disuse during the Roman Empire.

The primary method to give the illusion of a fully gold surface was to cover a base metal with a coating of thin gold sheets, called *foil* if thicker than about a micron (one thousandth of a millimeter), or *leaf* if thinner. The technique was used extensively in the Bronze and Iron ages. Foil usually was put into place and heated for the two metals to bond by solid state diffusion. Organic binders sometimes were employed. The most successful technique of attachment was *amalgamation,* used in China from about the third century B.C., and in the Mediterranean world from about the fifth century B.C. Gold was reacted with a small amount of mercury to form the amalgam, which was applied as a paste— as is done today in dentistry, particularly with silver. The base object became amalgamated with the paste to form a bond with the gold through solid state diffusion. Heating above the boiling point of mercury (356°C) then drove off the mercury and left pure gold behind as the surface. The Sassanians in Iran, from the second to sixth centuries A.D., produced many outstanding objects of gilded silver. The gilded silver plate in Figure 7.22 involved separate fabrication of the animals and placement on the plate, followed by gilding of the raised portions by amalgamation. This process was called *parcel-gilding* because some of the original silver surface remains. On rare occasions, gold was applied by dipping the entire object into a molten source, but the process was not easily controlled and hence was wasteful of gold.

Gold in the New World

The skill of the goldsmiths of pre-Columbian Peru and Colombia in South America has been rarely equaled. Most of the techniques were developed by

Figure 7.22. Sassanian silver plate with gilded figures, showing a king hunting ibexes, fifth or sixth century A.D. Reprinted with permission. © 1974 by the American Chemical Society.

the Moche Culture and passed on to their Chimú and Inca successors. Heather Lechtman has studied these techniques in detail. She found that South American goldsmiths may never have used amalgamation for gilding, and only occasionally did they dip the objects into molten gold or cover them with foil. Instead, they had an entirely novel arsenal of methods. One involved an electroplating process, whereby gold was dissolved in acid and then plated out onto a copper-containing object immersed in the solution. Dissolution of elemental gold, for example, by hydrochloric acid and other materials, produces a solution of gold chloride ($AuCl_3$). When a copper object comes in contact with the dissolved gold, the copper metal on the surface is unstable. Copper goes into solution as copper chloride, and gold plates out to maintain charge balance (Equation 7.9).

$$2AuCl_3 + 3Cu \longrightarrow 2Au + 3CuCl_2 \qquad (7.9)$$

Other methods took advantage of the novel alloy composed of gold, copper, and sometimes silver, called *tumbaga* by the Europeans. Many of the so-called gold objects from South America in fact were tumbaga with a gold surface. In order to obtain his release from the Spaniard Pizarro in 1533, the Inca king Atahualpa agreed to fill his prison room with gold objects up to the level to which his hand reached. After he did so, he was executed nonetheless, but on melting down the objects the Spaniards were greatly disappointed at the high level of baser elements. Many of the objects were expertly gilded.

Because of the low reactivity of gold, the copper and silver in the surface of tumbaga could be removed chemically, a process called *depletion gilding* by Lechtman. By one method, the copper/gold alloy was heated in air to convert metallic copper to its oxide, which was removed by a process called *pickling*. Copper oxide was dissolved off when treated with stale urine or acidic fruit juices, leaving a gold surface. The procedure also was used with copper/silver alloys with as little as 20 percent silver to impart a silvery appearance to the surface. With mixed gold/silver/copper (tumbaga), the surface could vary from silvery white to pale yellow, depending on the ratio of gold to silver. Lechtman described the possibility that a cementation process related to parting could have been used, whereby a paste of iron(III) sulfate and sodium chloride was applied and heated to form silver and copper chlorides, leaving behind the gold surface. All these techniques required subsequent surface burnishing. The Chimú and Inca thus had considerable control over the surface color of their objects, so that one must rely on chemical analysis before concluding that any product such as that in Plate 16 is in fact truly of gold.

Silver in Antiquity

Through much of the Old World Bronze Age, gold actually was a more common metal than silver, which is rarely found in the native state. Silver often accompanies gold in ores and could be obtained by parting, but more commonly silver was found along with lead ores. In fact, many of the lead mining sources mentioned previously may have been exploited primarily for their silver rather than their lead content, including the Aegean island of Siphnos prior to the fifth century B.C., Laurion from about that time, and Spain during the Roman period. The history of silver goes back, however, to at least the fifth millennium B.C., when small objects such as beads or buttons were produced in Iran and Anatolia. A fourth-millennium B.C. site at Byblos in Lebanon yielded more than two hundred silver objects. By the third millennium B.C., silver had become an important unit of exchange in Iran, Mesopotamia, and Anatolia. A Mesopotamian vase from about 2800 B.C. was produced from an ingot by repeated hammering and annealing and then was decorated (chased) by impressing a tool into its surface. Some evidence for the working of native silver is found in Hopewell sites in North America.

The most common ores of silver were found along with iron pyrite (FeS_2) or, more commonly, galena (PbS). Extraction from the ore was a two-step process. First the silver-containing lead ore was smelted. Roasting yielded the oxides, which were reduced by charcoal to the metals. The silver was separated from the argentiferous lead by a process called *cupellation,* named after the vessel or cupel in which it was carried out. The mixed metals were heated to about 1000°C in this open vessel of ceramic or bone ash, while a strong current of air was blown across it. The baser lead was oxidized to litharge (PbO) and absorbed by the vessel, leaving the unreacted silver, possibly mixed with gold. The gold and silver could be parted by cementation. The discovery of cupellation, which may have occurred as early as 4000 B.C., permitted the extraction of silver from very low-grade lead ore, down to 200 parts per million silver. The best deposits at Laurion contained only 4 kilograms of silver per ton of ore, about 0.4 percent. The litharge could be refined to produce lead. Silver from cupellation always contains a signature presence of lead, from 0.05 to 2 percent, so it may be distinguished from native silver and from silver that came from gold ore.

Although Egypt was rich in sources of gold, it had very little silver except as impurities in gold. It is likely that their gold was used as trade material in return for silver. That the trade was active is seen from the data of Stós-Fertner and Gale in Figure 7.23, which is a three-way, or ternary, plot of gold (Au), silver (Ag), and copper (Cu) from Egyptian artifacts prior to 1000 B.C. (see Figure 4.5 for a description of ternary plots). The gold artifacts frequently had 5–20 percent silver, and some had up to 26 percent copper, like tumbaga. When the silver content was 20–50 percent, the surface color was sufficient for the alloy to be called electrum. Analysis for lead showed that almost all the silver objects (those over 85 percent Ag) contained at least 0.1 percent and as much as 6 percent Pb, suggesting that the silver had been purified by cupellation, which was unknown to the Egyptians. Moreover, comparison of lead isotope ratios for these artifacts found no correspondence with isotopes from local Egyptian sources. Some corresponded to the Laurion field. The authors concluded that high-grade silver came from non-Egyptian sources.

The Sassanian people in Iran, from the second to the sixth centuries A.D., produced many extraordinary silver objects, such as that in Figure 7.22. Many objects were decorated on two sides, such as the bowl in Figure 7.24. Early workers, including some forgers, thought that they were produced by a double shell technique, whereby two pieces were decorated separately and then shaped together into the bowl and attached by bending the edges of the exterior piece over the interior piece. X-ray radiography by Pieter Meyers found only one piece. Apparently the skillful Sassanians had punched the depressions into the inner surface of a single piece, smoothed out the outer surface by scraping, and then decorated the outer surface.

Silver and gold artifacts sometimes were decorated by coatings of other metallic elements, analogous to enameling, which, however, is a ceramic. The substance *niello* (from the Latin *nigellum opus,* meaning "black work") was used since at least the Eighteenth Egyptian Dynasty, from which a nielloed ax and dagger have been found. Nielloed objects are very common from the Roman period, and Pliny, as usual, described instructions for their production. The active black ingredient was acanthite (silver sulfide, Ag_2S) or stromeyerite (mixed silver and copper sulfides, $Ag_2S \cdot Cu_2S$), possibly produced by fusing elemental copper, silver, and sulfur and mixing the result with a wax. The silver or gold object was incised to create a design, and the paste was used

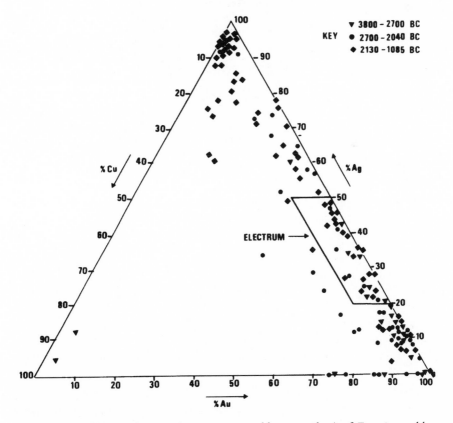

Figure 7.23. Ternary diagram (copper versus gold versus silver) of Egyptian gold objects.

Figure 7.24. Gilded Sassanian silver bowl, view of inside (bottom) *and outside* (top). *Reprinted with permission.* © *1978 by the American Chemical Society.*

to fill the depressions. The process of melting out the wax bonded the niello to the surface to create the striking black decoration.

Ancient Coinage

After pottery shards, coins are among the most common artifacts found in many excavations. Their metal content shows not only something about the history of metal usage, but also mirrors the sometimes rocky economies of the coin producers. Gold coins have been traced to Lydia (about 550 B.C.), but silver coins may have predated them slightly in Greece (about 580 B.C.). Greek silver coins contained 93–99 percent silver.

The process of debasement from high gold or silver to high copper is a story repeated throughout human history. As early as 474–450 B.C. in Sicily, the Syracusan tyrants increased the concentration of copper in their coins shortly before their expulsion. Fifth-century B.C. Macedonia may have maintained two parallel coinages. For domestic consumption a coin with a riderless horse contained 5–20 percent copper, whereas for foreign trade they used a coin with a horse bearing a rider and containing less than 0.2 percent copper. In the seventh century A.D., during the Dark Ages, Merovingian Gaul and the Anglo-Saxon kingdoms moved from gold coinage to increasingly higher silver content until the coinage was entirely silver, over about a fifty-year period. There may have been a scarcity of gold, or rising demand may have resulted in coins of lower face value. One of the most infamous events associated with devaluation occurred in England in A.D. 1124 when Henry I had the right hands and testicles of his minters cut off because they had allegedly debased the currency. Comparison of the silver content of coins from the period 1087 to 1125 found no difference by modern analysis from those for the period 1050 to 1087, averaging 93 percent. Thus whatever problems were bothering Henry I must have involved more general economic recession, and his draconian response was more in search of a scapegoat. Later in English history an actual debasement did occur during the civil war of the seventeenth century, when the king's gold coinage from Oxford dropped to 85 percent gold while the Parliament's London coinage remained at 91 percent (22 carat based on the maximum of 24).

A gradual debasement of coins is associated with the decline of the Roman Empire. The troubles actually began during the Republic, from the high cost of the Punic Wars. From the earliest times the basic Roman monetary unit was the *as,* composed of copper alloyed with tin and lead. It was divided into twelve *unciae* (from which the English words *inch* and *ounce* both come). During the third and second centuries B.C., the as dropped in weight from one pound (one *libra,* or lb., containing twelve unciae) to three unciae and then to one uncia. When Julius Caesar was murdered in 44 B.C., the as was valued at only a third of an uncia, or about 3 percent of its original value and weight.

In 15 B.C. Augustus reformed Roman coinage, making the silver *denarius* weigh $\frac{1}{84}$ of the Roman libra (about 325 g), the gold *aureus* $\frac{1}{42}$ of the libra, and the yellow brass *sestertius* equal to four copper asses. As Figure 7.25 shows, the reform was short-lived. Nero (emperor, A.D. 54–68) devalued the denarius by about 7 percent, and it continued to be debased until Domitian reversed the trend somewhat in 81. By the time of Marcus Aurelius it was down to 77 percent silver, and by the time of Septimius Severus, to 60 percent. Caracalla in 211 introduced the *antoninianus,* also called the *radiate* because it showed the emperor wearing a rayed crown. It weighed half again as much as the denarius, but by then both had only 50 percent silver. Efforts to retain the Roman provinces in Spain, Britain, and Gaul while holding off barbarian attacks required a vast military investment and eventually devastated the coinage. The denarius had essentially disappeared from circulation by the reign of

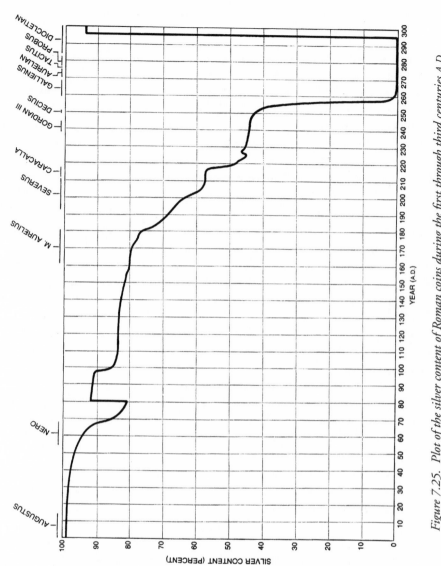

Figure 7.25. Plot of the silver content of Roman coins during the first through third centuries A.D.

Gordian III (238–244). By the mid-260s the radiate contained only 5 percent silver; and by 270, only about 1 percent: It had become a silver-dipped copper coin. From 200 to 400, however, gold coins tended to retain their content (though rarely circulating, like the American fifty-cent piece today, which is found mainly in Las Vegas), suggesting that there was a scarcity of the metal silver. Crawford estimated that during the early Roman Empire, about seventy-three tons of silver had to be added to the economy just to make up for wastage due to loss, hoarding, and so on. Diocletian's reform in 301 (Figure 7.25) was short-lived. The value of silver had simply outgrown its ability to serve as an object of commerce in small metallic units, exactly as occurred in the United States when coinage silver (92.5 percent) was abandoned in favor of cupro-nickel content (with a silverlike surface color) in 1976. Silver coins disappeared almost immediately from circulation, although the United States mint still makes them in denominations from ten to fifty cents in a noncirculating Silver Proof Coin, somewhat analogous to the rare high gold aureus of the late Roman Empire.

Elemental levels in coins provide not only a grim reflection of the economics of the time but also sometimes an indication of the refining techniques. Figure 7.26 plots the gold impurity in Sassanian silver *drachms,* studied by Adon Gordus. The level of gold was very close to 0.5–1.0 percent from A.D. 200 to 550. After A.D. 550, methods of refining led to much lower levels of gold. The open squares represent the gold content of known forgeries, and without exception they contain less than 0.3 percent gold and reflect the purity of silver produced in the nineteenth and twentieth centuries. The

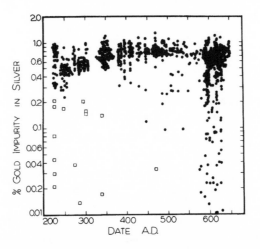

Figure 7.26. Gold impurity in silver Sassanian coins. Modern forgeries are shown as open squares. *Reprinted with permission.* © *1974 by the American Chemical Society.*

perceptive forger today must add a small amount of gold to the silver to achieve a match to the ancient alloy.

The Other Elements

It is often said that eight pure elements were known to the ancients, all metals: copper, tin, lead, zinc, iron, gold, silver, and mercury. The story of the one element not thoroughly considered thus far, mercury, goes back to the second millennium B.C. Heinrich Schliemann found a vessel containing mercury at Kurna from the fifteenth century B.C. A Late Bronze Age fitting may have been made of mercury-gilt bronze, found at Rathgall in County Wicklow, Ireland. The Chinese certainly had mercury by the Warring States Period (468–221 B.C.). According to legend the first Chinese emperor Chi placed a stratigraphic map of his empire in his tomb, complete with major rivers that were composed of mercury.

Mercury is mined as the pigment cinnabar (HgS) and is refined very simply because of its low boiling point (357°C). The ore is roasted in air above the decomposition point of cinnabar, and mercury vapors are collected by condensation. Mercury has been called quicksilver because of its color and liquid state at room temperature. The primary early use of mercury probably was as an alloying metal with gold and silver to form amalgams, both for application of these metals to the surface of other metals and for reclaiming gold. The Romans burned their used garments that contained gold thread and extracted the gold from the ash with mercury.

The ancients probably worked with more than the traditional eight pure elements. They possibly knew antimony, which could conceivably be smelted from the cosmetic stibnite. (Antimony sulfide, Sb_2S_3, was used by Egyptian women as an eye shadow.) A vase from the Chaldean site of Telloh dating to the fourth millennium B.C. appears to be made of antimony and small amounts of iron. The Romans and Greeks may have produced free antimony but failed to recognize it as being different from lead, which also is soft and dark gray.

Platinum was mined in pre-Columbian South America and was recognized as being a distinct substance from silver. It was probably found as the native metal in alluvial sands. An Egyptian box from the seventh century B.C. is said to contain an inlay of platinum, although the analysis dates to 1901 and is uncertain. Platinum became known to Europeans from South American samples only in the eighteenth century.

From the beginning of the human exploitation of metals, the new materials were appreciated for their hardness, for their ability to be formed into new shapes and to be repaired, and in some cases for their beauty. Their use led to expanded trade, craft specialization, harsher wars, and pollution, but generally to an improvement in life. The need for metals drove humans to search for better technology, particularly in winning metals from ore by smelting and in creating alloys with improved properties. This process has continued and even accelerated, as the electrical properties of metals were appreciated during the nineteenth century and the usefulness of newly discovered elements was exploited during the twentieth (aluminum for its light weight, titanium as an alloy with iron to form a hard steel, and platinum as a catalyst to speed up chemical reactions). The traditional metals, iron and copper, increasingly are being replaced by lighter and less corroding metals or even by organic polymers or composite materials such as polymers containing silicon carbide. The human appreciation of gold and silver continues unabated, so that it is unlikely that these metals will pass from our culture, even as the utilitarian metals of antiquity wane.

8

HUMANS

Temples and treasures may be exciting, but nothing rivals the discovery of ancient human remains: King Tut, the Ice Man, the Piltdown Man, and ancient bodies from peat bogs. These are the subjects that seize the public's imagination, and they too may be examined by chemistry.

The chemical legacy of an extinct people is present not only in the surviving creations of their hands (their artifacts), but also in their bodily remains, and even in the genetic material they pass on to their descendants. These human residues can provide a variety of information. The materials that we eat, drink, and breathe are metabolically transformed during our lifetimes into chemical markers related to our diet and health. When the markers survive in soil or bone, they may be analyzed by the chemist. Changes in other chemical markers from human remains can indicate the amount of time elapsed since burial and hence can aid in constructing archaeological chronologies. The molecules that make up genetic material change over many generations through natural mutations. Chemical analysis of these molecules and their mutations in living people as well as in archaeological residues is beginning to sort out human lineages and document prehistoric human migrations.

Diet and Isotopes

The most powerful technique for analyzing ancient human diet involves measuring, usually in bone or teeth, the relative amounts of the isotopes of elements commonly found in food. Carbon, for example, is the most important element in organics such as food, and occurs as two stable (nonradioactive) isotopes: ^{12}C, representing about 99 percent of the element, and ^{13}C, representing the remainder. Carbon dioxide (CO_2) in the stratosphere provides the ultimate source for carbon and is consumed by plants during the process of photosynthesis. From CO_2, plants assemble larger molecules, carbohydrates in particular, by forming carbon–carbon bonds. The more nimble, lighter isotope (^{12}C) forms these bonds faster, so that the resulting molecules are slightly richer in ^{12}C than is the atmosphere. Such a process is called *isotopic fractionation*.

Metabolism, however, is a complex process, and various plants may make a particular molecule by different sets of chemical transformations. The extent of isotopic fractionation, which determines the ratio of ^{12}C and ^{13}C as molecules are assembled, then, varies from plant to plant. Fortunately, there are only three major chemical pathways, and one of them is not particularly important in the study of ancient diet. All trees, all woody shrubs, and most grasses that grow in temperate or shady environments convert CO_2 initially to molecules related to glycerol that contain three carbon atoms. This pathway and the class of plants that uses it have been labeled C_3 (for three carbon atoms). Grasses from the subtropics (high temperature and sunny conditions) convert CO_2 to molecules with four carbons and hence have been labeled C_4. The third class includes cacti and succulents, which rarely contribute to diet.

There is a significant difference between the ratio of ^{13}C to ^{12}C in C_3 and C_4 plants. The observed ratio in any plant is compared with the ratio in an accepted standard—carbonate from a particular geological formation. Because the differences are small, they are expressed as parts per thousand (represented by the symbol ‰, read as "per mil" just as % is read as "per cent," respectively from the Latin words for "thousand," *mille*, and "hundred," *centum*). Scientists in general use the Greek letter delta (δ) to represent differences. In this context $\delta^{13}C$ is the symbol for the difference in the $^{13}C/^{12}C$ ratio between a plant and the standard, in units of ‰.

Dietary analysis by study of carbon isotopes is based on the observation of large differences in the $\delta^{13}C$ values between C_3 and C_4 plants, which are translated into similar differences in animal or human tissue. The foliage of C_3 plants on average has a $\delta^{13}C$ value of -26‰, and the value for C_4 plants averages -12‰. For reasons that are not entirely clear, consumption of plants by animals adds about $+5$‰ to the $\delta^{13}C$ in the organic constituent of bone called collagen. The kudu is an African animal that eats leaves from woody plants, and its collagen has a $\delta^{13}C$ of -21‰, characteristic of consumption of C_3 plants ($-26 + 5$) . Animals that graze the C_4 grasses have much less negative values, such as -7‰ ($-12 + 5$) for the tsesabe. Mixed feeders such as the sable antelope (-14‰) have intermediate values. These results are for plant eaters, or herbivores. Animals that prey on the herbivores (carnivores) may have slightly less negative values of $\delta^{13}C$.

The above results for modern herbivores show that dietary differences may be detected by isotopic analysis of carbon from bone collagen, as summarized by the following table.

Diet	$\delta^{13}C$ in the plant source	$\delta^{13}C$ in the consumer animal
Leaves from woody shrubs (C_3)	−26	−21
Mixed diet		intermediate
Subtropical grasses (C_4)	−12	−7

Humans are mixed eaters, or omnivores. Their diet is almost always a complex mix of components, and carbon isotopic analysis has added considerably to knowledge that previously was based on the observation of animal and plant remains at a human habitation site. Nikolaas van der Merwe, who pioneered the field of isotopic analysis, pointed out that this archaeological record, broadly speaking, is the menu from which the humans chose, but the isotopic record in bone helps to demonstrate what they actually consumed.

One of van der Merwe's most dramatic observations concerned the introduction of maize (corn) horticulture to the Woodlands of North America. Essentially all foodstuffs in this region originally were C_3 (woody shrubs, temperate grasses), until maize was introduced from Mesoamerica. As seen in Figure 8.1, the value of $\delta^{13}C$ in bone collagen hovered around the C_3 value of −21‰ from the Archaic Period (ending about 600 B.C.) and through the Early (600–150 B.C.) and Middle (150 B.C.–A.D. 250) Woodland periods, indicating a diet based largely on woody shrubs and temperate grasses. Sometime after A.D. 800, $\delta^{13}C$ started to become less negative, moving to about −18‰ by the end of the Late Woodland Period (A.D. 250–1000), to about −14‰ in 1200, and finally to almost −10‰ in 1300, during the Mississippian Period

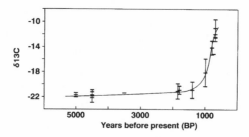

Figure 8.1. Changes in the proportion of carbon-13, expressed as parts per thousand, in human skeletal collagen from the North American Woodlands.

(A.D. 1000–1300). Maize, originally a tropical plant, uses the C_4 pathway to build organic molecules. As humans consumed increased amounts of maize, $\delta^{13}C$ in their bone became less negative, tending toward the value of −7‰ for a diet of only C_4 plants. If −21‰ is taken as indicating a diet devoid of maize and −7‰ a diet of 100 percent maize, the diet in A.D. 1000 represented about 24 percent maize; that in 1200, 45 percent; and that in 1300, 70 percent. These increases in the use of a cultivated food permitted large increases in population. Disease patterns that may be discerned from the bone, however, also increased, possibly as a result of insufficient dietary variability.

Similar changes occurred elsewhere in the Americas, but at different times. At Parmana in Venezuela, $\delta^{13}C$ changed from −26.0‰ in 800 B.C. to −10.3‰ in A.D. 400. These values support the theory by Anna Roosevelt and others that traditional cultivation of C_3 plants such as the cassava could not support the observed large increase in populations that occurred during this period. The gallery forests of the Lower Orinoco Valley lack C_4 plants, so the change in $\delta^{13}C$ requires a change in food sources. Introduction of maize cultivation, then, is a likely explanation for the change in $\delta^{13}C$. The archaeological record supports this interpretation, as flint chips used to grate cassava changed to grindstones to process grain. The early value of −26‰ is unusually large. Van der Merwe suggested that isotopic fractionation is increased by carbon recycling in the dense Amazonian forests. The same carbon atoms move from plant to leaf litter to CO_2 and back to the plant without thorough mixing with general atmospheric CO_2. If this cycle occurs repeatedly, $\delta^{13}C$ slowly becomes more negative. Analysis of leaf litter produced a value of −31‰, so that the prefarming value of −26‰ for the humans as usual represents a change of about +5‰ from the source food.

The carbon cycle in the oceans is more complex than that on land but must be understood in order to study cultures subsisting on marine products. The bottom of the food chain is plankton, then molluscs, zooplankton, and so on. Mollusc meat has a $\delta^{13}C$ of about −18‰, and predatory ocean fish and seals, of about −12‰. Thus the range is rather small. Van der Merwe compared the diets of two early burial groups from the southwestern Cape Province of South Africa for the period 2,000–4,000 years before the present. Coastal skeletons exhibited a mean $\delta^{13}C$ of −13.5‰, whereas inland skeletons had a mean of −18.9‰. He found that local marine-based foods had a mean of −15.6‰ and that terrestrial plant and animal foods had a mean of about −25‰. It is not surprising that the coastal population ate more marine foods (lower $\delta^{13}C$) than did the inland people. Because the two groups had different values of $\delta^{13}C$, the results eliminate an interpretation that there was only a single population that migrated seasonally between the two locations. The carbon isotope ratio in bones is stable for years and could not have changed over the course of a seasonal migration. Thus the two burial groups represent distinct populations.

Complexities arise when C_3 and C_4 terrestrial foods must be compared with marine foods. Michael J. DeNiro developed the use of nitrogen isotopes to provide another dimension to this method. Nitrogen occurs as two stable isotopes, the more abundant ^{14}N (99.7 percent) and the less abundant ^{15}N (0.3 percent). Metabolic fractionation of these isotopes occurs in the same way as with carbon isotopes. The standard of comparison is atmospheric nitrogen, for which $\delta^{15}N$ is set at 0‰. Legumes fix nitrogen directly without any isotopic fractionation and hence have a $\delta^{15}N$ close to zero. Other plants have a slightly higher value, about 3–10‰. Animals that feed on nonleguminous plants are higher (9–13‰), and carnivores that feed on these herbivores are a bit higher still (13–16‰), reflecting a food-chain effect. In the ocean the food chain is even longer and values can range from zero for air-fixing blue-green algae to 3–10‰ for marine plants and 15–20‰ for predatory seals and swordfish. These numbers are summarized below.

Diet	$\delta^{15}N$ in the plant source	$\delta^{15}N$ in the consumer animal
Terrestrial legumes	0–3	4–7
Terrestrial nonlegumes	3–10	9–16
Marine algae	0	unavailable
Marine plants	3–10	10–20

Thus very high values of $\delta^{15}N$ indicate that the diet contains marine products. Unfortunately, enhanced values of $\delta^{15}N$ also have been observed in very dry inland regions, also up to 20‰ for animals at the top of the food chain. Metabolic recycling of urea by animals under a water deficit may increase fractionation and enhance these values when protein is scarce in arid environments. Thus, $\delta^{15}N$ cannot distinguish marine from terrestrial foodstuffs in dry coastal regions. Further refinements are required for reef-based environments, in which the rate of nitrogen fixation tends to increase and cause lower values of $\delta^{15}N$.

It is common now to combine carbon and nitrogen isotope studies and to present the result on a two-dimensional plot (Figure 8.2). For terrestrial systems, distinctions are seen between herbivores and carnivores, between consumers of C_3 and C_4 plants, and between consumers of legumes and nonlegumes. For marine systems, distinctions are seen between herbivores and carnivores and between reef and nonreef fish. The darkened triangle simulates the range of isotopic values expected for a hypothetical feeding strategy for human inhabitants of a tropical island who consume maize (a C_4 plant), bivalves from a coral reef, and terrestrial carnivores that had fed on C_3 plant eaters.

By reference to Figure 8.2 and the above tables, the diet may be inferred for Neolithic Europeans, whose $\delta^{13}C$ of about −20‰ and $\delta^{15}N$ of about +8‰ indicate low marine food and high dependence on C_3 plants. The $\delta^{13}C$ and $\delta^{15}N$ values of −13 and +18‰ for Eskimo and Haida people (northwestern North America) indicate their much higher dependence on marine food. The respective values of −14 and +11‰ for prehistoric inhabitants of the Bahamas are consistent with consumption of foods from a coral reef environment, and the values of −7 and +9‰ for the inhabitants of prehistoric Tehuacán in Mexico indicate consumption of maize.

These isotopic ratios all were measured on the organic collagen component of the skeleton. Bone tends to resist decomposition better than other parts of

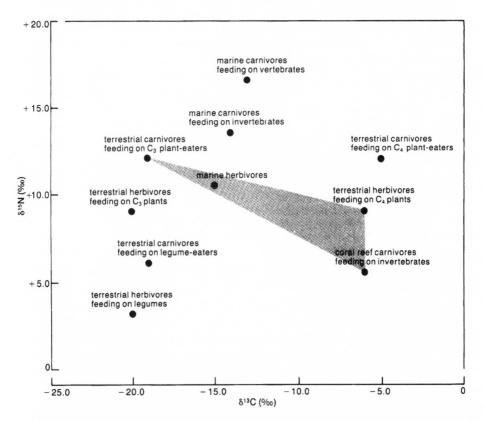

Figure 8.2. Comparison of the proportions of carbon and nitrogen isotopes expected in bone collagen for humans feeding on specific types of food (dots). The triangle represents the region of isotopic proportions for a mixed feeding strategy for the diets at the three vertices.

the body, so that it is more widely available. Carbon, of course, could also be found in soft tissue such as skin or hair; however, this tends to decay more rapidly. The inorganic portion of bone or teeth that provides their support function consists of the molecule hydroxyapatite, with the formula $Ca_{10}(PO_4)_6(OH)_2$. Some carbonate (CO_3^{2-}) can substitute for phosphate (PO_4^{3-}) and provide an alternative source of carbon for isotopic analysis. Moreover, the inorganic part of bone remains much longer, up to millions of years, whereas collagen usually decomposes completely within ten thousand years. Study of isotope ratios in apatite thus can extend the range of the method. Moreover, carbon in apatite comes from different biochemical sources than the carbon in collagen. Stanley H. Ambrose has demonstrated that collagen primarily reflects protein intake, whereas apatite carbonate reflects nonprotein intake (grasses and shrubs). Measurement of $\delta^{13}C$ in both organic and inorganic parts of bone may provide more detailed information on diet.

The multilevel approach of examining carbon and nitrogen isotopes in both collagen and carbonate promises to help unravel dietary complexities, including relative contributions of proteins and carbohydrates, of marine and terrestrial resources, and of wild and cultivated plants. In addition, isotopic analysis may reveal dietary differences in populational subgroups based on gender, social status, and age.

Diet and Trace Elements

The positive ion calcium (Ca^{2+}) in hydroxyapatite may be substituted by various metals present in the diet. In 1965 H. Toots and M. R. Voorhies pointed out that replacement of calcium by strontium (Sr^{2+}) may be related to the position of the organism on the food chain, or the *trophic level*. At the lower end of the food chain, plants absorb Sr (strontium) from groundwater at essentially the geological background level. When herbivores consume plants, they have a natural biochemical preference for Ca (calcium) over Sr during the formation of bone. The Sr level then is lower in herbivore bone than in plants. Moreover, very little Sr is present in flesh, so that carnivores consume much lower levels of Sr than do herbivores, and even less Sr is present in their bone.

Thus levels of Sr decrease with position on the food chain: Sr is highest in herbivores, then omnivores such as humans, and finally carnivores. Consequently measurement of Sr concentration in bone can help place the animal on the food chain (trophic level). Arnold L. Rheingold used modern roadkills to confirm this relationship, finding 400–500 ppm of Sr in the bone of herbivores, 150–400 ppm in omnivores, and 100–300 ppm in carnivores (ppm is parts per million; 10,000 ppm equals 1 percent). James H. Burton has argued that Sr levels reflect primarily plant components in the diet rather than meat or trophic levels in general.

This very attractive approach to determining diet or trophic levels of animals suffers from a serious drawback when put into practice with archaeological samples. Unlike the carbon of collagen, inorganic elements in hydroxyapatite can exchange with the burial environment, a process called *diagenesis.* A soluble element can wash or leach out, particularly in a wet environment. Elements from the soil can be absorbed by the bone. The extent of uptake and loss of elements depends on the type of soil, the water table, the elevation of the site, the pH of the soil (acid or base), the soil temperature, the levels of elements preexisting in the soil, and the presence of collagen-degrading microorganisms.

Thus the Sr level in bone can increase or decrease during burial, rendering conclusions about trophic levels inaccurate. The same would be true of any other element substituting for Ca that might have trophic significance. Efforts have been made to remove nonbiological Sr from buried bone, based on the expectation that contaminating Sr is more soluble than the original biological Sr. (Newly introduced Sr is weakly bound, but biological Sr is integrated thoroughly into the apatite.) Andrew Sillen has developed a method using successive washings that remove contaminating (diagenetic) Sr while retaining the original (biogenic) Sr.

Another way around this problem is to compare groupings within populations, which should share a common degree of contamination. Burials often contain information about the status of the individual, such as grave goods or placement within a cemetery. Margaret Schoeninger found that the highest-status individuals in the Chalcatzingo site in Mexico, as indicated by the presence of jade objects, had the lowest levels of Sr. Studies by our group of the Middle Woodland Gibson site in Illinois found that the highest-status individuals, as judged by location in the burial mound, had the lowest levels of Sr. In both cases it was thought that lower levels of Sr indicated access to better food, which contained higher levels of protein, for example from meat.

T. Douglas Price and his co-workers have found that barium (Ba) is even more sensitive to trophic levels than Sr. The series Ca, Sr, and Ba are all alkaline earth elements, from the second column in the periodic table. The atoms increase in size from Ca to Sr to Ba. Replacement of Ca by Sr deforms the bone matrix, but Ba is larger than Sr and causes even greater deformations. For this reason the body discriminates against Ba more than Sr, giving rise to larger concentration changes based on diet.

It is problematic whether other elements can be used in dietary analysis. They must be largely resistant to exchange during burial, and they must be subject to little physiological control during life, which could mask dietary effects. Nonessential elements such as Sr and Ba thus are ideal, since the body's primary response is discrimination. Zinc (Zn) and sodium (Na), on the other hand, are essential elements and are subject to physiological control in a number of ways. Nonetheless Zn has been found empirically to provide some trophic distinctions. It works in the opposite direction to strontium. In his roadkills, Rheingold found 90–150 ppm of Zn in herbivores, 120–220 ppm in omnivores, and 175–250 ppm in carnivores (Zn occurs in relatively high levels in meat). Despite difficulties in trace element analysis for understanding diet, it appears that levels of Sr and Ba are useful when proper care is taken to reduce the effects of contamination.

Lead and Health

Elemental analysis of bone also can provide information on pathological conditions. Chapter 7 described how the excessive use of lead-based materials by Roman populations resulted in very high levels of lead (Pb) in human skeletons. As with other elements, there are difficulties in separating the true effects of lead from those caused by contamination or leaching. Lead is absorbed by the body from food through the intestine, from air through respiration, and even by simple contact through the skin. Because there is little natural exposure to lead, our distant ancestors did not evolve a process for its efficient excretion. Consequently, excessive exposure results in accumulation, almost entirely in the bone. It takes about twenty years after deposition for the bone to lose half its lead burden through normal reworking. Thus continued exposure results in further accumulation.

Arthur C. Aufderheide has examined the lead levels of several archaeological populations. He compared two distinct burial groups at the Clifts Plantation site in Westmoreland County, Virginia (A.D. 1670–1730). Just as the effects of strontium were minimized by comparing groups of individuals with similar burial conditions on the basis of status, Aufderheide minimized the effects of lead contamination by comparing two groups with similar burial conditions. One contained the skeletons of five planters and another contained eleven laborers. Class distinctions were based on grave goods. The average Pb content of the skeletons of the planters was 185 ppm, with a range of 128–258 ppm. That of the laborers was 35 ppm, with a range of 8–96 ppm. For comparison, ancient Peruvians (who used no lead-containing materials) had 1 ppm or less of Pb, and a modern American has about 50 ppm of Pb in the skeleton, according to the work of Jonathan Ericson. The Clifts coffins were of pine, not lead, so that the differences did not result from lead-containing coffins. Aufderheide suggested that the planters used upscale pewter and lead-glazed earthenware vessels from which lead could leach.

Aufderheide also studied slave populations from the sugar plantations of Barbados in the cemetery of the Newton site over the period from 1660 to 1820 (emancipation occurred in 1834). In forty-eight individuals he found an average of 118 ppm, which is more than three times the levels in the laborer population of Clifts Plantation. Since the Barbados slaves did not have access to pewter products, Aufderheide attributed their high levels of lead to the consumption of rum, which was distilled through equipment made of lead. Alternatively, sugar manufacture itself may have involved lead products.

K. J. Reinhard and A. M. Ghazi found high levels of lead in Omaha Native American burials dating after contact with Europeans. They attributed it to the use of metal materials such as ornaments or musket balls. They carried the analysis a step further by comparing the Pb isotope ratios of skeletal Pb with that in the artifacts. Their similarity supported the conclusion that the high skeletal levels were caused by the European trade goods.

Attempts have been made to relate levels of iron to pathologies such as cribia orbitalia, a bone lesion in the orbital roof in the eye. Unfortunately, iron is a major component of soil, along with manganese and aluminum, and it easily contaminates bone. No study has succeeded in separating the pathological and contaminative effects of iron.

Nitrogen and Fluorine Dating

Elemental contamination is not all bad. In fact, it has been exploited very effectively to draw chronological conclusions about human bone. The introduction of unnatural elements or the depletion of natural elements takes place over time. For example, fluorine (F) is present in modern bone at the level of 0.03 percent or less. Exposure of buried bone to groundwater naturally containing fluoride ions results in exchange of the hydroxide (HO^-) component of hydroxyapatite to form fluorapatite, $Ca_{10}(PO_4)_6F_2$. This latter material is extremely stable and becomes fixed in bone, building up slowly with time. Geologist Kenneth P. Oakley found 0.3 percent of F in a Neolithic skull from Coldrum, Kent, and 1.7 percent of F in the early Paleolithic Swanscombe skull. The increased levels of fluorine thus provide a crude measure of the age of the material. The appearance of uranium (U), either by radioactive decay of other elements or by environmental contamination, can provide a similar measure, as modern bone contains no uranium. Oakley found the Swanscombe skull to contain 27 ppm of U_3O_8.

The element nitrogen (N) is a natural component of collagen in bone and is lost as collagen hydrolyzes slowly during burial. Modern bone contains about 5 percent N, which dropped to 1.9 percent for Oakley's Neolithic skull and was found only in trace amounts in the Swanscombe skull. Thus loss of N is another measure of age.

The uptake or loss of these elements is strongly dependent on the burial environment. As a result, these elemental methods are not able to provide an absolute date, but often they can provide relative chronology within a single site in which burial conditions are essentially constant. Nitrogen and fluorine dating thus have been used to develop chronologies of burials with uncertain stratigraphy. Clusters of burials may have a confused chronology when later burials intrude upon earlier burials, or if bones were removed without recording the original stratigraphy, as can happen at construction sites. Although radiocarbon methods can give reliable absolute dates from small amounts of collagen, the ease of elemental analysis can provide a large survey of skeletal material. Moreover, if the bones have been treated with an organic preservative or if they are of a great enough age to have lost all their collagen, carbon dating is not possible.

M. Marchbanks used both F and N dating to determine the sequence of twenty-four burials at the Blue Bayou site in Victoria County, Texas. The results indicate that various areas of the cemetery in fact were used at different times, with more recent burials in the southern part and older burials in the northern part. A. Haddy and A. Hanson carried out a similar study on human bones excavated during the 1930s at the Moundville site in Alabama. These samples had been treated with polyvinyl acetate and hence could not be dated by carbon methods. Based on pottery classification, this site has three periods: Moundville I (A.D. 1100–1250), Moundville II (1250–1400), and Moundville III (1400–1550). They analyzed fifteen samples and obtained reliable N and F percentages on eleven of them. The results are compared in Figure 8.3, and a rough negative correlation was observed (F goes up as N comes down). The relative dates from N and F percentages correlated very well with available chronology based on pottery.

Possibly the most famous use of elemental dating was in the case of the Piltdown skull. This well-known hoax lasted from the first report of the bones in 1912 until Oakley measured the levels of N and F in 1950. The perpetrator is commonly thought to be Charles Dawson, the county solicitor and amateur

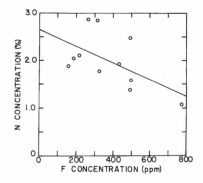

Figure 8.3. Comparison of nitrogen and fluorine levels in eleven bone samples.

archaeologist who allegedly found the skull in a gravel bed on Barkham Manor, near the village of Piltdown, Sussex. Early human remains had been reported on the Continent, particularly the Heidelberg jaw in 1907, and fossil hunters were actively searching for further examples of archaic humans. In 1912 Dawson informed a friend, geologist Arthur Smith Woodward, that he had discovered fragments of a human skull at Piltdown. Later in the year, joined by theologian and anthropologist Pierre Teilhard de Chardin, they found further skull fragments and an elephant molar. Reconstructions of the skull were made, and in 1913 Teilhard found a canine tooth that fitted the most apelike reconstruction. The dissimilarity of the cranium and the jaw was recognized very early. In 1914 William King Gregory of the American Museum of Natural History stated that "they may represent a deliberate hoax, a Negro or Australian skull and a broken ape jaw, artificially fossilized." In 1915, however, a second Piltdown discovery was made by Dawson (left frontal bone, occipital bone, molar tooth). Before he revealed the location of these discoveries, he died in 1916. Gregory reversed his position when confronted with these samples.

The discovery of human fossils in China by Teilhard and W. C. Pei and in South Africa by Raymond Dart during the 1920s showed that the Piltdown remains were clearly anomalous. Piltdown became a side issue as better documented finds came to light. The final blow was provided by Oakley, and every fossil associated with Piltdown was revealed to have been faked. The Piltdown skull had 0.1 percent F and 1.4 percent N, suggesting slight antiquity but not as early as the late Pleistocene. The jawbone had <0.03 percent F and 3.9 percent N, both indicative of a thoroughly modern bone. Thus the skull and jaw were not of the same date and did not have the antiquity required for the apparent geological context. The jawbone is now thought to be that of an orangutan, and the skull that of a modern human, just as Gregory had suggested. The bones had been treated with potassium dichromate, iron, or manganese to simulate age. The canine tooth found by Teilhard was from a modern ape and had been filed. The elephant molar had 2.7 percent F and no N and hence was of considerable antiquity, possibly early Pleistocene. Thus the perpetrator had gone to considerable lengths to provide plausible material.

Dawson is the most likely perpetrator, as he had ample opportunity to place all the finds, he had a known history of shady activities, and he was eager to obtain greater respect in professional paleontology circles. There is no lack, however, of other conspiracy theories. Stephen Jay Gould accuses Teilhard of being directly responsible, but he had little motive. Others consider that it was someone else's bad joke taken seriously by the professionals. In Agatha Christie fashion, Keith Stewart Thomson and Leonard Harrison Matthews have devised complicated schemes involving multiple individuals operating independently. They suggested the role of a disgruntled naturalist from the British Museum (Natural History) named Martin A.C. Hinton. He had access to bone raw materials and knowledge of chemical treatment. Moreover, he had a wage dispute with the geologist Woodward. Indeed, the contents of his trunk were discovered recently in a museum loft and reported by Brian Gardiner to contain bone samples treated chemically as the Piltdown materials had been. Hinton, however, lived until 1961 and could have treated these bones at any time after the hoax to determine whether chemistry could simulate the appearance of burial. In the final analysis, it is the hoax itself that piques the interest of the public, rather than who actually did the deed.

Organic Molecules from Bodies

Bone is only one component of the human body, which contains a myriad of organic and inorganic molecules ranging in size from carbon dioxide and nitric oxide to enormous proteins and nucleic acids. The dietary and chronological studies just described were not concerned with the individual identity of molecules, so long as they derived originally from the human body and not from the environment (other than F and U dating). Specific identification of molecules leads to entirely new information, but the questions remain as to whether the molecules changed after death and whether the molecules actually came from the human. One thus must examine the chemistry of death, or thanatochemistry. The action of bacteria, fungi, water, and air can bring about a variety of oxidation, reduction, hydrolysis, and other reactions. Detailed molecular analysis thus is required to sort out molecules present during life from those resulting from degradation or contamination after death.

Richard P. Evershed and colleagues have pioneered the study of organic remnants from the past. Chapter 6 has mentioned their work with food residues. They also have examined extracts from ancient human bone. Their studies with modern bone established which small organic molecules are present. Their preliminary work with ancient bone has endeavored to identify molecules that come from the body, as distinguished from those introduced from the environment. Archaeological conclusions remain to be developed.

Evershed and co-workers found that the molecule most closely associated with the bone matrix itself is cholesterol. Moreover, cholesterol is not normally present in soil and hence is unlikely to be a contaminant. Their procedure was to crush the bone, extract the organic components with solvent, and separate them by gas chromatography (GC). Figure 8.4 shows the resulting gas chromatogram for a human tibia from the late Roman or early Saxon period in Britain, A.D. 300–500. The largest peak in the chromatogram is from cholesterol, whose structure is given in the left insert. The peak labeled IS3 is an internal standard added to the bone to permit quantification of components. By comparison of the sizes of the cholesterol and IS3 peaks, they estimated that about 0.00001 grams of cholesterol was present in each gram of dry bone. The next largest peak is from a close molecular relative of cholesterol, in which a ketone group (C=O) has been added at the position labeled number 7. As this material was not present in samples of modern bone, it must be a result of a chemical reaction after death. The fact that it is an oxidation rather than a reduction product (oxygen is added to, not removed from, the molecule) indicates that the burial conditions were oxidative. The oxidation agent could be either atmospheric oxygen or enzymes provided by microorganisms. This work demonstrated that organic molecules from life survive burial.

Evershed found a richer organic yield from the skin of several ancient bodies taken from peat bogs in the Netherlands. Collagen in skin, bone, teeth, and muscle tissue is naturally tanned by the peat bog, which is anaerobic (lacking oxygen) and contains tannins or tanninlike materials that can strengthen the polymer and enhance preservation. Like the process of mummification, peat

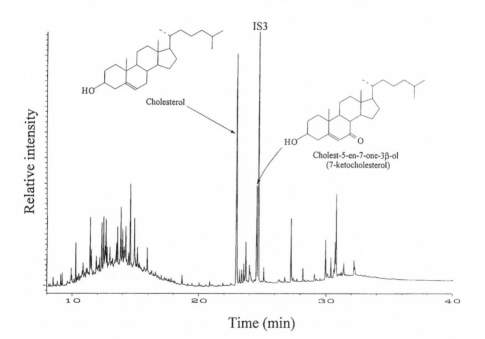

Figure 8.4. Gas chromatogram of lipids extracted from a fourth-to-sixth century human tibia from the Saxon period in Britain. Each vertical peak represents a distinct compound, two of which have been indicated with structures. The peak labeled IS3 is a standard for comparison.

bogs largely preserve only the outward appearance of the body. Extensive chemical reworking occurs. Evershed's analysis found no compounds called phospholipids, which make up the major structural components of the living cell membrane. By microscopy, in fact, the cellular structure was completely destroyed. Only the collagen fibers of the original skin survived as the apparent bog body skin.

Figure 8.5 shows the GC traces from the extracts of the skin of three Dutch bog bodies—one from a girl's foot (A) and two from a man's leg (B and C)—radiocarbon dated to about 30 B.C. They had been found around the turn of the twentieth century in Yde and Weerdinge. The peaks marked 1 to 13 and 18 are from long-chain fatty acids (R-CO_2H, in which R is a long hydrocarbon chain). Oleic acid is the major fatty acid found naturally in living human skin, but it was entirely absent. Its degradation product, 10-hydroxyoctadecanoic acid, formed by adding a molecule of water to the double bond, was detected as the large component 7. Another component (15) is cholesterol. In contrast to the Roman bone, the Dutch bog skin lacked the oxidation product, 7-ketocholesterol. Instead, two major reduction products (14 and 16, respectively called 5β- and 5α-cholestan-3β-ol) were found. The 5β form, also called coprostanol, occurs naturally in living intestines by microbial action, but the 5α form rarely occurs naturally and was likely produced by a degradative reduction.

Figure 8.6 summarizes the chemical relationships and shows a possible intermediate (17) between cholesterol (15) and coprostanol (14) that also is present in the chromatogram. Phytosterols such as sitosterol (19) are abundant in peat and presumably are contaminants. Chemical analysis of the bog bodies thus confirmed that hydration and reduction occur under burial conditions, but that small molecules from the body survive long after cell structure is gone. These organic molecules repel water and hence are said to be hydrophobic: this property contributes to their stability in the aqueous peat environment.

Fossilized Excrement

Scientists, eager to obtain information about diet, also have analyzed the excrement of ancient humans. Petrified dung, called *coprolites*, contains molecular and biological components that indicate what the human ate and hence provides a direct measure of the diet rather than of the menu. Coprolites usually are found in extremely dry locations, and caves in deserts have been a rich source. Some from the Lovelock Cave in Utah are 3 inches in diameter. They are softened by soaking in a dilute solution of sodium phosphate (Na_3PO_4). The process also extracts much of the bile pigments. The softened coprolite is broken up and passed through graduated sieves to recover biological residues

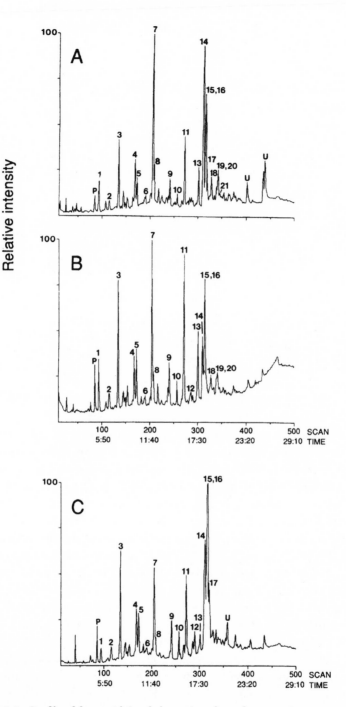

Figure 8.5. Profile of fatty acid (methyl esters) and sterols present in extracts of bog body skins, as measured by gas chromatography and mass spectrometry. The samples were from Yde (A, skin from a girl's foot, 30 ± 80 B.C.), Weerdinge (B, skin from a man's leg, not dated), and Weerdinge (C, skin from a man's leg, 30 ± 70 B.C.), all in the Netherlands.

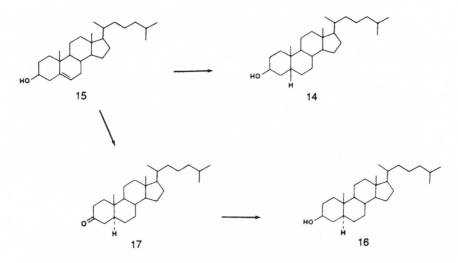

Figure 8.6. Chemical relationships between cholesterol (15) *and its degradation products found in lipid extracts of bog body skins: 5ß-cholestanol (*14, *also called coprostanol), 5α-cholestano-3-one* (17), *and 5α-cholestanol* (16).

such as seeds and fibrous debris. Sometimes flotation methods with benzene and salt solutions also are used to separate components.

Richard S. MacNeish studied a long, chronologically defined series of coprolites from caves near Ocampo in southwestern Tamaulipas, Mexico. From the earliest period (7000–5000 B.C.) the coprolites showed that prickly pear cactus *(Opuntia)* and runner bean were consumed. The fruit, or tuna, of the prickly pear is a common, nutritious, and tasty food eaten today in Mexico and parts of the United States. By 2300 B.C., chili peppers, pumpkin, and agave had been added (agave is a large-leafed succulent that also provides alcoholic beverages and cordage). By 1800 B.C. the coprolites also contained residues from sunflower and aloe and later mesquite. Few residues of maize appeared, even after evidence for its farming was secure, possibly because maize kernels were thoroughly ground into flour before consumption.

Robert F. Heizer studied the rich lode of coprolites from Lovelock Cave and found remains from plants (bulrush and cattail), fish (*Siphateles* or chub), insects (the water tiger beetle), aquatic birds, and rodents. From the excavation, it appeared that living areas also served as the latrine. The sameness and relatively unnutritious nature of the diet bespoke a very hard existence.

Chemical analysis of coprolites (scatochemistry) also has been carried out. The presence of coprostanol (14 in Figure 8.6), already noted as a natural decomposition product of cholesterol, has been found to be a reliable molecular marker for fecal pollution (it has been used in studies of environmental pollution of marine and lake sediments). Evershed has examined archaeological soils for the presence of coprostanol to study issues such as waste disposal, manuring, diet, and health. In one study, a feature at a Roman site in Stanwick, Northamptonshire, U.K., was thought to be a cesspit, and the appearance of the soil suggested two separate periods of use. Figure 8.7 contains a diagram of the feature. Samples were taken from the upper and lower portions of each layer (samples 1–4) and from undisturbed sterile soil outside the feature (samples 5 and 6). Neither the undisturbed soil nor the upper portions of the two layers contained coprostanol, but both lower layers did (samples 2 and 4). The chemical analysis thus proved that the feature was a cesspit and that there were two episodes of use separated by infill or disuse.

Amino Acid Dating

Earlier in this chapter, the organic portion of bone was used both as a carbon source for isotopic analysis of diet and as a nitrogen source for dating. The material, collagen, is an example of the large class of molecules found in the body called *proteins*. Structural proteins such as collagen contribute to the overall form of the organism and make up in part bone, muscle, skin, nails, and hair. Catalytic proteins, called enzymes, control and accelerate chemical reactions that occur in a living organism. In an archaeological sample, natural processes degrade these proteins over time, so that only highly stable materials, such as collagen in bone and skin or keratin in nails and hair, remain. The large protein molecules are called polymers because they are made up of small units (monomers) that are bonded together repetitively. For the case of proteins, the monomer units are amino acids that usually have the formula $NH_2\text{-CHR-}CO_2H$, in which R can have a variety of structures. A protein may have dozens or hundreds of amino acid units.

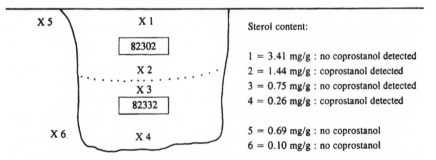

Sterol content:

1 = 3.41 mg/g : no coprostanol detected
2 = 1.44 mg/g : coprostanol detected
3 = 0.75 mg/g : no coprostanol detected
4 = 0.26 mg/g : coprostanol detected

5 = 0.69 mg/g : no coprostanol
6 = 0.10 mg/g : no coprostanol

Figure 8.7. Diagram of a cross section through a feature suspected to be a Roman cesspit from Stanwick, Northamptonshire, U.K. Six samples (X1–X6) were analyzed to show two distinct layers separated by a fecally uncontaminated region.

These amino acids may be recovered from archaeological samples and analyzed in various ways. In particular, determination of the arrangement of the four groups around the central carbon atom (H, R, NH_2, CO_2H) in three dimensions has led to a controversial dating technique. Nature has optimized chemical reactions in biological systems by making the reacting atoms fit together most comfortably in space. Spatial arrangements of atoms are referred to as *stereochemistry.* Nature has carried stereochemical optimization to the point of even distinguishing between mirror images of a molecule. When the mirror images are identical, no distinction of course can be made, but when they are different (or nonsuperimposable), one mirror image often can react more rapidly and hence provide more efficient biochemical reactions.

Amino acids are examples of molecules with nonsuperimposable mirror images, as seen in the following diagram:

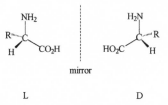

L   D

The form on the left is called an L (for *levo*) amino acid and that on the right a D (for *dextro*) amino acid, based on conventions concerning the arrangement of the four groups (NH_2, CO_2H, R, and H) around the central carbon atom. For reasons that have never been understood, the appearance of life on earth resulted in the preference of the L forms as the building blocks for almost all proteins. Thus the D forms usually are associated with nonliving materials. They are found on meteorites, and they appear in dead organisms over time through natural chemical reactions involving, for example, removal of the hydrogen atom and replacement on the opposite side of the molecule.

Philip H. Abelson suggested that the appearance of D amino acids in proteins of dead organisms should be able to serve as a means of determining how long ago death had occurred. Jeffrey L. Bada has been instrumental in developing this chemical process into a practical dating technique. Because the process of converting an excess of L amino acids (or of D, for that matter) into a 50/50 mixture is called racemization (oddly, from *racemus,* Latin for "bunch of grapes"), the technique is called *amino acid racemization dating.*

The experiment is superficially simple, as it is necessary only to measure the relative amounts of the D and L forms of an amino acid widely present in proteins, a measurement that has been done since the nineteenth century. Bada selected the amino acid aspartic acid (R– is $HO_2CCH_2–$ in the above formulas) for racemization studies, as it is abundant and has a convenient half-life of about 15,000 years in bone at 20°C (during this time a sample moves halfway from pure L to the 50/50 mixture). To determine how fast aspartic acid racemizes under archaeological conditions, Bada measured the D/L ratio for aspartic acid in a bone from the Olduvai Gorge in Tanzania with a known radiocarbon date of 17,500 years. Subsequent measurements of aspartic acid racemization in samples from Iran, Hungary, and Mallorca gave dates calculated from the Olduvai calibration that were in excellent agreement with known radiocarbon dates. The calibration process was intended to allow for the fact that racemization is highly dependent on the temperature, acidity, and water content of the burial site.

Problems arose when the method was applied to Paleoindian skeletons from California. The amino acid dates of 30,000 to 70,000 years were far beyond the likely arrival of humans to North America. Either there was a serious flaw in the amino acid method, or North American chronologies needed to be revised radically. R. E. Taylor subsequently radiocarbon dated the California samples and found that they all dated to a period of about 2,000 to 8,000 years ago, ages that fitted comfortably into accepted chronologies. The method of amino acid dating thus needed to be reassessed.

Taylor enumerated several factors that can contribute to inaccuracies in the amino acid method. Most come down to the difficulty of assuming that the sample used for calibrating the racemization process experienced the same conditions as the sample to be dated. The temperature can vary, as can the water content of the soil and bacterial and fungal levels that hasten decomposition. Taylor found that six individuals from a single site in California (Baldwin Hills) had D/L ratios that varied from 0.12 to 0.49, giving racemization dates ranging from 2,800 to 48,000 years, compared to the radiocarbon date range of 4,000 to 5,300 years before the present. Taylor noted that higher racemization was found in more decomposed bones, as measured by nitrogen levels (the sample with D/L = 0.12 had 1.1 percent N, but the one with D/L = 0.49 had only 0.0022 percent N), indicating that racemization rates depended on the extent of decomposition. If this observation is general, it represents a serious flaw in the method. More fully decomposed bones would always exhibit exaggerated ages.

The most optimistic interpretation of amino acid dating of bone is that it should give accurate dates for well-preserved samples when calibration conditions closely parallel those for an unknown sample. Recent experiments on eggshells, particularly of ostrich, emu, and extinct ratite, suggest that these materials are much less subject to diagenesis than bone, because of the compact nature of the calcite matrix. Amino acid dating thus has experienced a renaissance and has seen extensive use in dating African sites that are older than the limit of radiocarbon methods and that contain eggshell residues.

Amino acid racemization also has been used with some success for an entirely different objective, estimating the age of an individual at death. The high physiological temperature of mammals (37°C) accelerates the process of racemization. D-amino acids do not build up in most proteins during life because they are excreted after normal rebuilding of tissue. For the case of tooth enamel and dentine, however, tissue does not turn over rapidly, and D-amino acids accumulate with age. When a person dies or a tooth is extracted, amino acids no longer are exposed to the high physiological temperatures, and D-amino acids cease to build up. The level of D-aspartic acid then serves as a measure of the age at death or at extraction of the tooth. This technique can apply only to individuals who died relatively recently (possibly up to a thousand years). After longer times, D-amino acids would build up even at the lower nonphysiological temperatures of burial, in the manner originally used by Bada for dating the sample.

For modern tooth extractions, A. M. Pollard and his group found a steady linear relationship between the level of D-aspartic acid and the age at which the individual lost the tooth: about 1.1 percent for age 10, 2.0 percent for age 28, and 3.0 percent for age 50. Results with excavated samples were not so good. Burials at Spitalfields, London, (eighteenth and nineteenth centuries) gave only approximate agreement between racemization ages and actual ages at death from the tombstones. The results ranged from quite good (within 2 to 5 years of the actual age) to significant overestimates (a measured age of 45 for a 19-year-old). Figure 8.8 shows a plot of the percent of D amino acid versus the age at death (or loss of the tooth). The blackened circles are the values from the modern calibration group, and the open circles are the Spitalfields results. Except for a group of outliers, the Spitalfields samples fall nicely on the calibration points, but the age is usually overestimated. Although the technique is useful only for relatively recent samples, it is a promising approach for establishing age at death from teeth alone.

Dating by Electron Spin Resonance and Thermoluminescence

One of the desired advantages of amino acid dating was based on the longer half-life of aspartic acid racemization than that of carbon-14 radioactive decay. It was hoped that amino acid techniques could provide dates beyond the upper limit to carbon dating of about fifty thousand years. Another chemical technique, based on the analysis of free electrons, has found widespread use in the dating of older samples. Normally electrons are paired in chemical bonds or orbitals. A free, or unpaired, electron can be found in organic radicals, transition metals, and some common small molecules such as oxygen.

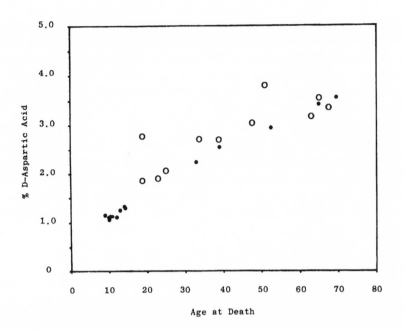

Figure 8.8. Plot of the amount of D-aspartic acid from collagen in the first premolars versus age of the individual at the time of death or tooth loss. Modern teeth from individuals of known age are the filled *circles, and samples from Spitalfields burials are the* open *circles.*

Unpaired electrons are studied by the technique known as *electron spin resonance* (ESR), also called electron paramagnetic resonance (EPR). The sample is placed in a magnetic field. The unpaired electrons act like little bar magnets that interact with the field. Application of a small amount of microwave energy can alter this interaction and result in the absorption of energy by the sample. An ESR spectrum is composed of a plot of the amount of energy absorbed versus the value of the magnetic field. A peak in the spectrum consequently signifies the presence of an unpaired electron.

For dating purposes, it is necessary that living tissue such as bone or teeth have essentially no unpaired electrons. Over periods of time on the order of ten thousand to more than a million years, natural radioactivity generates unpaired electrons in bone and teeth. The energy of radiation from radioactive uranium (U), thorium (Th), or potassium (K) is sufficient to knock electrons, in low concentration, out of their stable orbitals. For the most part these free electrons quickly return to their orbitals, but occasionally they are trapped in a defect in the crystal lattice. These defects arise when the wrong atom gets into the structure (like aluminum in place of calcium) and disrupts the normally well-ordered crystal. Some traps are stable enough to retain the unpaired electrons over these long time periods, so that the number of unpaired electrons slowly builds up. The ESR experiment measures the number of unpaired electrons directly. The larger the signal, the older the sample. The trapped electrons may be released if the sample is heated strongly. Thus it is important that bone not have been subjected to heating between deposition and analysis.

Other materials, such as pottery, brick, and various types of stone, also can contain trapped electrons that build up over time. Whereas for bone this buildup begins at death, for these other materials it begins when they were fired. Artificial heating in the laboratory also releases the free electrons. This process is accompanied by the emission of ordinary light and is called *thermoluminescence* (TL), a dating technique widely used for such materials. Unlike TL, which measures the release of the unpaired electrons and hence irreversibly alters the sample, ESR simply counts the number of unpaired electrons and leaves them in place.

For dating by either TL or ESR, it is necessary to determine how fast free electrons build up. For ESR, this calibration involves measuring how much uranium is present in the bone or tooth (Th and K are negligible), and how much of all three radioactive elements are present in the surrounding soil or water in which the object was found. The calibration is complicated when uranium actually moves from soil into the object and hence starts to have a stronger effect. Two models have been developed, depending on whether the majority of the uranium accumulated soon after burial (early uptake) or accumulation was continuous (linear uptake). ESR dating can be extended back about five million years and TL, 500,000 years; both techniques vastly exceeding the range of radiocarbon dating.

ESR dating has provided some very critical measurements in developing current theories of the evolution of human ancestors (hominids) over the last million years. The group of K. D. Sales, G. V. Robins, and D. Oduwole has studied paleolithic bones from the site of Zhoukoudian in China (previously called Choukoutien). These were not the famous Peking Man samples, which were lost mysteriously during World War II, but they were excavated in the same cave. Figure 8.9 shows what they call the dating signal, as other ESR signals from manganese or amino acid radicals also can be present but do not build up in the desired fashion. The dating signal must have been absent at the beginning of the time period to be dated, it must have increased over time from a steady dose of natural radioactivity, it must be stable for a much longer period than the age of the sample, and it must not have been affected by intermediate events. Such events could include a fire, restructuring of the crystals during natural diagenesis, and grinding during sample preparation. The scientists had to determine the natural dose of radioactivity. Given the reasonableness of this measurement, they measured an age of at least 260,000 to 570,000 years for these *Homo erectus* individuals.

R. Grün has carried out ESR studies at later stages of human evolution, when modern *Homo sapiens* was first appearing. The initial discoveries were in Europe and were referred to as Cro-Magnon people. They originally were considered to evolve from or replace Neanderthal people (called Neandertal in the current scientific literature, to parallel German spelling of *Tal*, for "valley"). Discoveries in Africa and Southwest Asia, however, have altered this scenario profoundly. Grün dated mammalian teeth associated with hominids found in the Jebel Irhoud cave in Morocco. These hominids had archaic features but resembled modern humans rather than Neanderthals. ESR dates of 90,000 to 190,000 years were obtained. Another sample from the eastern Sudan, essentially of a modern human with some archaic features, gave an ESR date of 97,000 to 160,000 years. The large ranges resulted from uncertainties in the dose rate of natural radioactivity. Samples of twenty individuals from Qafzeh cave, near Nazareth, Israel, appeared to be archaic *Homo sapiens* and gave consistent ESR dates of 100,000 to 120,000 years. ESR also has been used to date early Neanderthals, including a skeleton from La-Chapelle-aux-Saints in France (47,000–56,000 years), LeMoustier in France (40,000–45,000 years), Guattari Cave in Italy (44,000–74,000 years), and Kebara Cave in Israel (60,000–64,000 years).

Thus modern humans were in Africa long before they appeared in Europe, and early modern humans preceded or were contemporaneous with

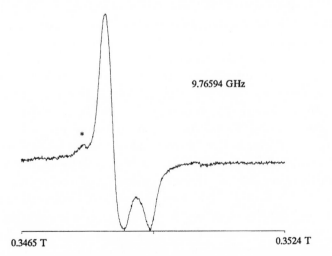

Figure 8.9. Electron spin resonance spectrum of a sample from Zhoukoudian bone. The large peaks are the dating signal and the small peak marked with an asterisk is extraneous.

Neanderthals. Neanderthals probably evolved in Europe from *Homo erectus* in response to the rigors of glaciation starting as early as 500,000 years ago. They may have been largely cut off from other *Homo* populations by geographical or climatological barriers. After modern humans developed in Africa about 100,000 years ago, the two populations must have overlapped in Europe and Southwest Asia. Neanderthals lasted in Spain up to about 30,000 years ago, when they were replaced by modern humans. It is still possible that Neanderthals contributed to the gene pool of modern humans because of the long overlap.

In addition to dating teeth and bone, ESR has been used for numerous other purposes, including determining the origin of marble, measuring the temperature of the heat treatment of flint and ivory, and understanding cooking practices with grain. Along with TL, ESR is the method of choice for the study of thermal histories.

Genetic History from Blood Analysis

The studies of Evershed described earlier in this chapter showed that some organic molecules can survive burial conditions, but the information available from them is still very sketchy. Population geneticists have developed an entirely different approach to analyzing organic molecules by reasoning from the present into the past. Many such molecules are genetically determined, so that their structures are passed on in an unchanged or definable pattern from generation to generation. Thus analysis of such molecules in present populations may provide information about past populations. Moreover, if modern populations can be associated historically with particular locations, their biomolecules can serve as models for specific earlier populations. In order to avoid systematic errors that may be associated with a single biomolecule, it is best to examine as many as possible. Because genes represent the most fundamental biomolecules in a historical study of people, this work usually involves analysis of genes or of their physical or chemical manifestations in modern populations.

In particular, information obtained from blood analysis has become central to such populational studies, because blood has been analyzed for decades, because analytical techniques based, for example, on immunology are both sensitive and accurate, and because many genetic markers are readily found in blood. This work was pioneered by Luigi Luca Cavalli-Sforza. Three examples suffice to illustrate his method. First, the ABO blood groupings vary considerably around the world. Native Americans are all of the O blood group, so that incursions of exotic populations can be recognized by the presence of other blood groups.

Second, the Rh negative gene is peculiarly European, or rather Caucasian. The highest frequency, more than 50 percent, is found among the Basques, who live near the northern Pyrenees mountains on the border of Spain and France. Their language is very different from the Indo-European languages spoken by most Europeans. Anthropologists have suggested that the Basques may represent a remnant of pre-farming (pre-Neolithic) Europeans and may be genetically closest to the Cro-Magnon inhabitants of Europe who were replaced by the people who brought farming from Southwest Asia and caused a rapid increase in population. Today the Basques are more than half Rh negative, whereas other European groups are 40 percent or less Rh negative. The Paleolithic Europeans may have been almost all Rh negative, whereas the Neolithic invaders were almost all Rh positive. European populations are still undergoing the process of conversion and eventually will become nearly all Rh positive. The isolation of the Basques has slowed the conversion of their population.

As the third example, the condition known as cystic fibrosis, which involves faulty digestion, breathing difficulties, and excessively salty sweat, is primarily genetic. It probably originated as a mutation involving the change of just one amino acid in a protein. Figure 8.10 shows the frequency of the mutant gene responsible for most cystic fibrosis conditions, called ΔF-508. The highest

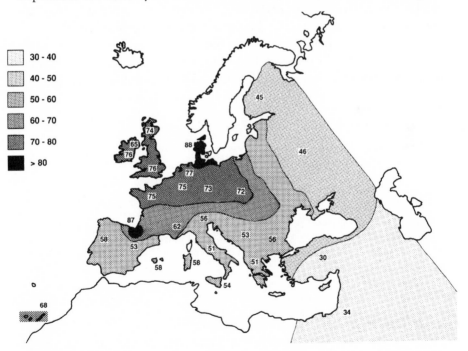

Figure 8.10. The relative frequency of the common mutant gene for cystic fibrosis (ΔF-508) in Europe, in comparison with other mutants for the disease. Reproduced with permission of Princeton University Press. © 1994 by the Princeton University Press.

concentrations are in northern Europe and around the northern Pyrenees and again may represent a characteristic of the pre-Neolithic population.

The program of genetic analysis from blood tests carried out by Cavalli-Sforza examined populations that have been least affected by the European expansion that followed Columbus's discovery of the New World. Although these groups may have been subject to many other movements of populations, such as the barbarian migrations that followed the fall of the Roman Empire and the Muslim expansion following the founding of Islam, they represent a reasonable measure of local genetic characteristics because of the stability of rural populations prior to the Industrial Revolution.

Cavalli-Sforza and colleagues concentrated on forty-two populations that had continuously occupied their pre-Columbian geographic locations. They originally studied thirty-eight genetic factors and later expanded the study to 120. These factors included ABO blood types, the Rh factor, other blood characteristics (HLA-A, HLA-B, MN, haptoglobin), and other genetic markers such as enzymes (phosphoglucomutase, acid phosphatase). Each type of factor occurs as two or more genetic variants, called *alleles,* such as blood types A, B, and O. The frequency of the entire set of alleles for the population groups was translated by statistical analysis into a mathematical figure called the *genetic distance,* which represents how different one group is from another genetically. The following table, from the work of Alberto Piazza, illustrates a simple analysis of Europeans based only on HLA (human leukocyte antigen) alleles:

	Danish	German	French	Austrian	Italian	Spanish	Turk	Basque	Sardinian
Lapp	585	596	637	612	741	846	954	837	1200
Danish		32	66	44	138	130	269	175	563
German			77	55	89	91	164	188	553
French				37	90	67	223	126	389
Austrian					65	92	210	134	411
Italian						47	82	184	360
Spanish							95	131	336
Turk								298	473
Basque									385

Read the names across from and above a given number to specify a genetic distance. The largest numbers (all of which should be multiplied by a thousand) show that the Lapps and the Sardinians exhibit the greatest differences from the others, followed by the Basques and the Turks. Only small differences in these genetic markers exist among the other groups, particularly the Germans and the Danes.

When all forty-two populations are treated in this fashion, based on 120 alleles, the resulting genetic distances can be represented by a linkage tree such as shown in Figure 8.11. Only thirty-eight groups are listed because the five European groups are collected together. The length of the horizontal lines represents the genetic distances when measured to a vertical connecting line. Thus the bottom two groups (New Guinean and aboriginal Australian) differ by about 0.016. Connections of these two groups with others require larger horizontal distances measured after moving along the vertical connector. Eventually all the groups are connected at some distant time. The earliest split (the vertical line furthest to the left) occurs at 0.030 between all the non-African groups and the six African groups (except for the Berbers, who resemble Caucasians),

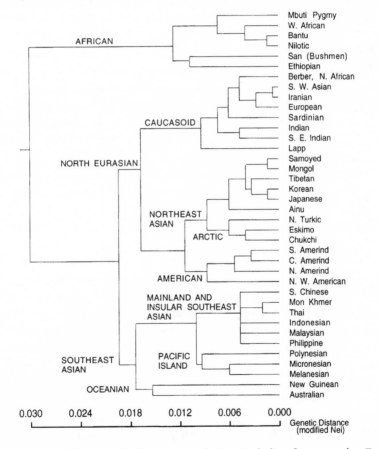

Figure 8.11. Linkage tree for forty-two populations, including five grouped as European. The horizontal coordinate represents genetic distances based on 120 allele frequencies. Reproduced with permission of Princeton University Press. © 1994 by the Princeton University Press.

meaning that Africans and non-Africans exhibit the greatest genetic differences. The non-Africans split first into two groups, which contain on the one hand the North Eurasians and on the other hand the Southeast Asians. Thus Japanese and Koreans are genetically closer to Native Americans and Europeans than to Malaysians and Polynesians.

These variations suggest routes for the populating of the world, which is illustrated in Figure 8.12, from the work of Cavalli-Sforza, Piazza, and also Paolo Menozzi. Based on archaeological evidence, they suggest a movement of *Homo sapiens* from their origin in Africa about 100,000 years ago, a peopling of Asia and Australia about 60,000 years ago, arrival in Europe about 35,000 years ago, and penetration into the Americas between 15,000 and 35,000 years ago. As populations separated, genetic mutations started to create differences: the longer the separation, the greater the number of genetic differences. Thus the non-Africans separated from Africans at the earliest stage, so these groups have the largest genetic distances seen in Figure 8.11, although later mixing resulted in variations such as the Berbers.

The 120 different alleles can be combined mathematically, since many of them behave similarly. The resulting combinations are principal components, which have been used in earlier chapters in numerous other multidimensional contexts. Figure 8.13 (p. 244) is a plot of the two most significant principal components. Separation of the world populations is seen from a different perspective, with Caucasians for the most part in the upper right, Africans in the lower right, Southeast Asians in the lower left, and Northeast Asians in the upper left.

Principal component analysis can be particularly edifying when superimposed on regional maps. Figure 8.14 (p. 245) is a plot of the first principal component for African populations. Cavalli-Sforza and his co-workers interpret these gradients as representing the north–south changes from a Caucasoid population to a Negroid population. Each principal component represents one or more distinct factors giving rise to genetic variation. Gradients in the fourth African principal component (not illustrated) may arise from a genetic expansion associated with a distinctive African agriculturalism emanating from Cameroon.

The first principal component map for Asia (Figure 8.15, p. 246) clearly shows an east–west gradient, which may have been associated with a wave of people or activities from northeastern Africa or Southwest Asia. The second principal component is a north–south gradient, expressing the differences between Mongoloid groups.

Finally, the first principal component for Europe (Figure 8.16, p. 246) has a southeast–northwest gradient that is associated with the introduction of farming from Southwest Asia. The figure closely resembles the genetic trend of the

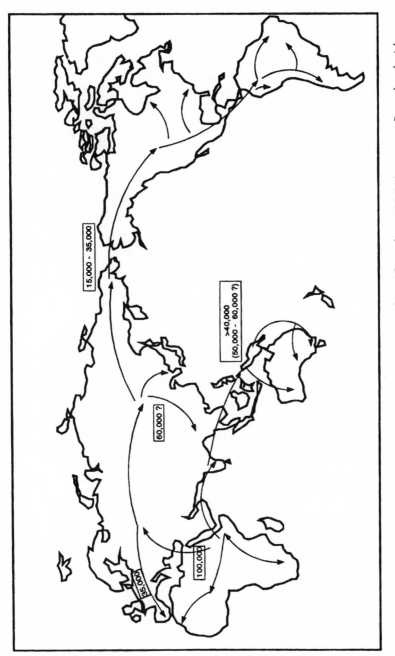

Figure 8.12. Possible migration routes for modern humans, starting from Africa about 100,000 years ago. Reproduced with permission of Princeton University Press. © 1994 by the Princeton University Press.

cystic fibrosis factor in Figure 8.10 and a map based on the archaeological dating of the arrival of farming across Europe (Figure 8.17, p. 247). At the times noted on the latter plot, the hunter–gatherer subsistence strategy gave way to Neolithic farming. As originated in Southwest Asia between 8000 and 7000 B.C., farming was based on wheat, barley, peas, lentils, and animal husbandry. The independent discovery of farming in northern China between 4500 and 3500 B.C. was based on foxtail millet cultivation and pig domestication, and in Mesoamerica around 6000 B.C. on maize, chili pepper, squash, and bean cultivation. These developments must have resulted in waves similar to that evident in the European map. The European farming revolution also may have brought the Indo-European language base.

Two mechanisms have been discussed by anthropologists for the changes in subsistence and language patterns during the European Neolithic wave. *Cultural diffusion* involves the transmission of techniques from group to group without the movement of people. *Demic,* or population, *diffusion* requires the

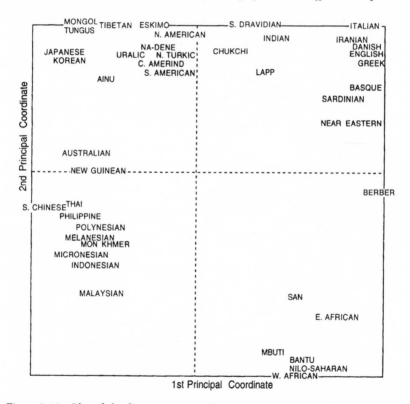

Figure 8.13. Plot of the first two principal components of allele frequencies for the forty-two human populations. Reproduced with permission of Princeton University Press. © 1994 by the Princeton University Press.

farmers themselves to move. The first principal component map of Figure 8.16 is presented by Cavalli-Sforza as strong evidence for the demic mechanism, as genes and hence people migrated. The introduction of farming vastly increased the carrying capacity of the land, so that populations increased throughout Europe with the arrival of farming. The spread of farming over Europe was very slow, about one kilometer a year. Cavalli-Sforza and his co-workers imagine a gradual absorption of the pre-Neolithic hunter–gatherers, followed by population increase. The Southwest Asian genes were increasingly diluted as farming progressed across Europe, resulting in the gene gradient seen in Figure 8.16.

The genetic history of Italy based on this approach indicates that much of today's populations can be traced back *in situ* to pre-Roman times. Figure 8.18 (p. 248) contains the maps for the first (left) and second (right) principal

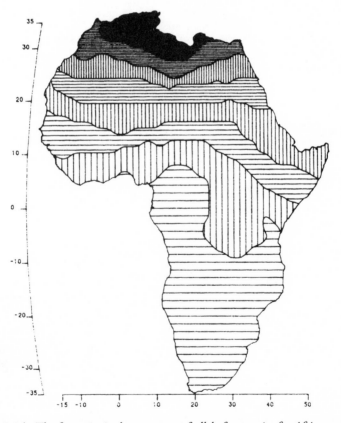

Figure 8.14. The first principal component of allele frequencies for African populations, plotted geographically. Reproduced with permission of Princeton University Press. © 1994 by the Princeton University Press.

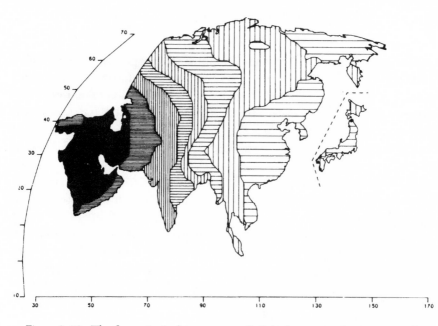

Figure 8.15. The first principal component of allele frequencies for Asian populations, plotted geographically. Reproduced with permission of Princeton University Press. © 1994 by the Princeton University Press.

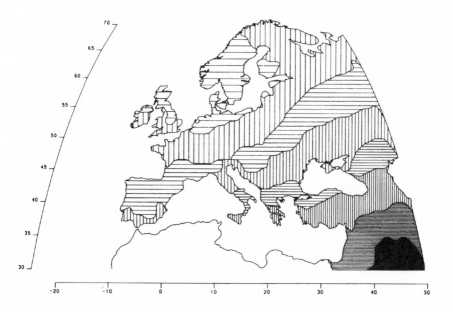

Figure 8.16. The first principal component of allele frequencies from European and western Asian populations, plotted geographically. Reproduced with permission of Princeton University Press. © 1994 by the Princeton University Press.

components. The north–south gradient of the first component results from the mix of northern Europeans with Mediterranean groups, including Greeks, during the period of Magna Graecia (from the ninth century B.C.) and Arabs (between the seventh and eleventh centuries A.D.). The east–west gradient in Sicily results from pre-Greek, Greek, Norman, and other settlements. The east–west differences also are seen in the second principal component. The most striking feature in this map, however, is the focus, just north of Rome, in southern Tuscany. This is the historical home of the Etruscans, and the feature may be a reflection of pre-Roman genes. One of the reasons that Italian genes may still represent ancient groups is the large populations in ancient times. The Italian population at the height of the Roman Empire has been estimated to have been six to seven million people, possibly the densest in Europe. After a decrease during the Middle Ages, the population was up to about ten million by 1300 A.D. These robust groups were able to maintain a strong genetic identity despite various invasions. Lower-population areas would be more liable to dilution or destruction by the arrival of other groups.

It is worth pausing for a moment to appreciate that analysis of these same blood molecules in ancient samples could be extremely valuable to the archaeologist. Cavalli-Sforza restricted his modern human groups to those that had a

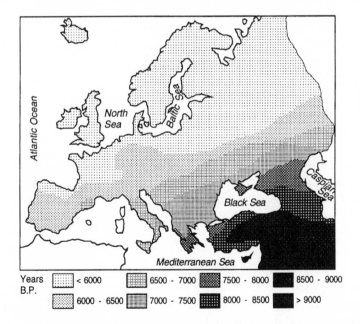

Figure 8.17. The expansion of farming from its source in Southwest Asia through Europe during the period 6,000 to 9,000 years before the present. Reproduced with permission of Princeton University Press. © 1994 by the Princeton University Press.

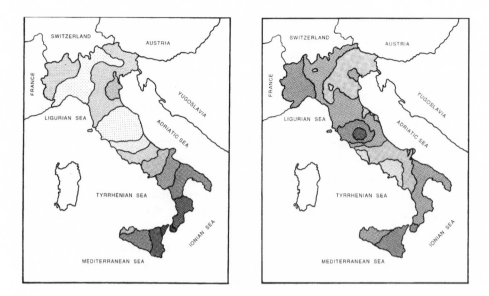

Figure 8.18. The first (left) *and second* (right) *principal components of allele frequencies for Italian populations plotted geographically. Sardinia is not included. Courtesy of the Cambridge University Press.*

demonstrable geographical location prior to the time of Columbus. If ancient samples could be analyzed by the same techniques, even deeper connections could be made between groups. Such experiments are thwarted by the threefold problems of availability, contamination, and degradation. Blood samples are difficult to find archaeologically and always occur in very small amounts. They may be contaminated by samples from other species or by similar molecules from other natural sources. The natural environment is constantly exposing ancient molecules to conditions that alter them chemically, such as oxidation and hydrolysis. Although these difficulties are substantial, it is possible that well-preserved samples can be analyzed for kinship and lineage information. To date, however, ancient blood has not been analyzed in this fashion.

Ancient and Modern DNA

In terms of kinship and lineage, the most valuable molecules should be those that encode genetic information, the *deoxyribonucleic acids* (DNA). Like other biological molecules, ancient DNA is subject to the problems of availability, contamination, and degradation. Its unique structure, however, has made possible the development of a method to avoid some of these difficulties. Like proteins, DNA is a linear biopolymer, that is, it is composed of a unit that is repeated over and over along a chain. The repeating unit in the nucleic acid chain is made up of a carbohydrate (the ribo part, also called a sugar) and a phosphate. An additional organic piece called a base is attached to the carbohydrate. These repeating units are quite different from the amino acid units of proteins.

The individual nucleic acid units vary only in the structure of the base attached to the carbohydrate. There are four possible bases, with the names adenine (A), guanine (G), cytosine (C), and thymine (T). These bases contain hydrogen atoms on nitrogen (N–H) that can loosely bond with atoms on other bases in what are called *hydrogen bonds*. In DNA, two entire chains are bound together by these hydrogen bonds in a tightly defined process in which A on one chain pairs only with T on another, and G only with C. Because one specific base of one chain is always hydrogen bonded to another specific base of the complementary chain, the paired groups are collectively called *base pairs*. The resulting structure is the now familiar coil or helix, with the two DNA chains bound like the opposite lanes of an expressway, although in a coil.

Because hydrogen bonds are very weak in comparison with normal chemical bonds (about a tenth the strength), the two parts of a DNA helix easily come apart and unwind during reproduction. One unwound strand then acts as a template for the biochemical synthesis of another complementary strand. In this way genetic characteristics embedded in the order of bases (A, G, C, T) on the DNA chain are passed on to a new generation without change (although mixed with the DNA of another individual in sexual reproduction).

The ability of paired DNA chains to become unpaired and to be replicated was exploited by Kary B. Mullis to develop a method to reproduce (or amplify) DNA in the laboratory. Using an enzyme called DNA polymerase, he was able to use unwound DNA molecules as templates to synthesize exact copies of portions of the strands and hence to increase the amount of available material. Called *polymerase chain reaction* (PCR), the method enables the molecular biologist to take a small amount of an archaeological sample and amplify it to a level at which it can be analyzed and characterized by standard chemical methods. In the same fashion as Cavalli-Sforza's blood properties, the DNA molecules can then help define a population.

In the archaeological context, the PCR method cannot alleviate the problem of degradation, but it has to a large extent solved the problems of availability and contamination. Only very small amounts of DNA need be found, and the PCR method can amplify them to a usable quantity. Starting with one DNA chain, thirty or forty repetitions of the PCR can produce more than a million identical chains. In addition, the highly technical PCR experiment utilizes primers to initiate the process of amplification. Primers are used that select the desired DNA molecule and ignore the contaminants. They can operate in the presence of overwhelming quantities of DNA impurities from foreign or highly degraded sources. The method thus is both sensitive and specific. Archaeologists and molecular biologists are now engaged in the necessary task of characterizing DNA from present and past populations.

Human tissue, particularly bone, yields a few micrograms (10^{-6} of a gram) of DNA from a gram of material. Excavated samples might contain only 1 percent of the amount expected from modern tissue. Contaminant DNA may come from the burial environment (such as peat), from the excavator, or from the molecular biologist involved in handling the tissue. The recovered material may have a few hundred or a thousand base pairs in the chain. It appears that DNA can survive for thousands of years under reasonable conditions, especially under cold, dry, or anaerobic conditions that encourage preservation of soft tissue in general. Preservation up to millions of years has been claimed.

There also is the question of which type of DNA to target. DNA is found both in the cellular nucleus and in the mitochondria (small bits of the cell that help produce energy). Mitochondrial DNA (called mtDNA) is a single, short nucleotide with only 16,569 base pairs for humans, compared with 6×10^9 in the entire library of some forty thousand DNA genes (the genome) in nuclear DNA. Moreover, there are up to ten thousand copies of mtDNA in a cell, whereas nuclear DNA has only one. Thus handling mtDNA appears to be easier. Also, each individual normally has only one kind of mtDNA, whereas nuclear or chromosomal DNA comes in two types, one from each parent. Mitochondrial DNA occurs as a single type because it is inherited unchanged only from the mother. It offers the advantage of encoding specific genetic information about the maternal line. This simplifying aspect has the disadvantage that genetic information about male contributions to population dynamics—such as from armies, hunting parties, or tribal hierarchies—is lost. It also is useful that mtDNA exhibits a great deal of variability in each base pair sequence, so that comparison between populations is simplified. Greater sequence similarity is expected the more closely two groups are related.

Finally, mtDNA mutates more rapidly than nuclear DNA, so that changes over time may be followed more easily with mtDNA. Mutations may be fatal and then lost to the population, they may be favorable and slowly build up in a population, or they may be neutral. Since mtDNA conveys no outward characteristics, all its mutations are neutral and carry no evolutionary advantage or disadvantage. It has been estimated that in the most variable part of mtDNA, 11.5–17.3 percent of the base pairs change by mutation over a million years.

The few archaeological studies of ancient DNA have endeavored to demonstrate the viability of the material over time and to establish groupings based on sequence differences. The group of William W. Hauswirth studied the large, well-preserved, and quite ancient human remains found at the Windover site in the central coastal region of Florida. Radiocarbon dating indicated an age of 7,000 to 8,100 years ago. During this thousand-year period, numerous burials took place. The pond conditions encouraged organic preservation of fabric, wood, and human soft tissue. The archaeologist in particular wanted to establish genetic continuity of the burial population. Analysis of both nuclear and mitochondrial DNA found persistence of specific types that suggested occupation by a single related population over the thousand years. They were able to compare their specific results with larger studies of modern Native North Americans. Several laboratories have characterized four classes of Native American mtDNA sequences, or *haplotypes* (called by some I, II, III, and IV, or by others, correspondingly B, D, A, and C). The archaeologists found that their population had sequences that were within the extremes of the other studies, so they could conclude that no non-native sequences were observed. About a third of the Windover samples fell into two of the four known sequence groups (I and II), but the remaining samples fell into no specific group. They suggested that further haplotypes of Native Americans must have existed. These later were found in modern Native populations and called X6 and X7.

Mark Q. Sutton and co-workers have examined DNA isolated from those hardy coprolites from the American western desert, one from Lovelock Cave in Utah (no date listed, other than "several thousand years ago"), and three from the open La Quinta site near Palm Springs, California (dating to about five hundred years ago). By isolating DNA and using PCR primers specific to X and Y chromosomes, they were able to demonstrate that one coprolite probably came from a male and two from females (the fourth yielded no recoverable human DNA). With the sex of the individual known, archaeologists now can relate diet, as determined by the classical biological methods described earlier in this chapter, with gender.

The origin of the first settlers on Easter Island and other islands in the eastern South Pacific has been controversial since the famous voyage of Thor Heyerdahl in the *Kon Tiki* from Peru to Tuamotu Island in the 1940s. Anthropologists generally agreed on a Polynesian origin of the Easter Islanders, but Heyerdahl favored a South American origin of the people who constructed the colossal heads. Erika Hagelberg has isolated mtDNA from the sites of Ahu Tepeu (A.D. 1100–1680) and Ahu Vinapu (1680–1868) on Easter Island and found it to resemble DNA from several Polynesian sources. Thus the first genetic evidence favors a Polynesian origin for these people. A larger controversy has arisen over whether these expert seafaring Polynesians reached South America and contributed to the gene pool of Native Americans. The higher proportion of B haplotypes in South America than in North America, and its common presence in Polynesia but not in Siberia, suggest that the B type entered the American gene pool via Polynesia. Anne Stone, however, found that an eight-thousand-year-old individual discovered in a cave in Colorado's White River National Forest has the B lineage. Since Polynesia was not settled until six thousand years ago (and the easternmost islands only a thousand years ago), the presence of the B haplotype in North America long preceded any possible incursion from Polynesia. This is the expected pattern from the traditional origin of the B type in Asia via the land bridge to Alaska, rather than in Polynesia. Thus no case can be made for connections between Polynesia and North America, in either direction, based on ancient DNA.

The group of Robert E. M. Hedges is involved in a program to characterize the genetic history of British populations. The results of Cavalli-Sforza from blood molecules demonstrated gradients that move from the south to the north, reflecting Viking invasions from the north, Saxon invasions from the south, Roman influences from the south, or possibly even Neolithic settlements from the Continent in pre-Roman times. As background for making comparisons between Romano-British populations and the Saxon invaders, Hedges carried out DNA surveys of modern Welsh populations (stand-ins for the ancient British) and German and Danish populations (Saxon stand-ins). They found a rich diversity of German and Danish haplotypes, but still almost a fifth were identical. Since this same haplotype was found among the Welsh, it appears to represent a very ancient population. Five of six samples from an Anglo-Saxon cemetery and one from a Roman-British cemetery also gave the same common sequence, as did a six-hundred-year-old sample from Hulton Abbey. Such results suggest that alternative parts of mtDNA are going to have to be examined in hopes of finding greater variability, or nuclear DNA will have to be studied. The analysis of ancient human DNA thus has much developmental work still ahead of it.

On the other hand, analysis of DNA in modern humans is now a straightforward experiment. With the help of PCR, it can be done on a strand of hair or other easily accessible body parts. The experiment is largely a statistical analysis of the differences between individuals' DNA. If a large-enough DNA segment is examined, there will be differences, as all humans except identical twins, are genetically distinct. The differences represent either the different genetic characteristics of the individual or neutral mutations.

During the 1980s the late Allan G. Wilson and his group pioneered the application of modern mtDNA to human lineages. The number of base-pair differences in the mtDNA of two individuals could be interpreted in terms of how long ago they had a common ancestor, if the mutation rate is known. Wilson originally used the archaeological data for the initial settlement of New Guinea (30,000 years ago), Australia (40,000) and the New World (12,000), and the mtDNA differences of these groups from the rest of the world to calculate that 2–4 percent of the base pairs mutate every million years. They found that human mtDNA on the average varies by about 0.57 percent between individuals. Consequently, convergence occurred 140,000 to 290,000 years ago (respectively, 0.57 percent/4 percent and 0.57 percent/2 percent multiplied by one million years for these limits). At this time, either all humans had the same mtDNA, or only one type of mtDNA existing at that time survived in later populations.

Figure 8.19 illustrates this process. It considers six different hypothetical mtDNA haplotypes *(a–f)*. Individuals *e* and *f* have very similar mtDNA, so their common ancestor is recent (very few mutations have occurred). Greater dissimilarities of *e* and *f* from *b, c,* and d mean that their common ancestor is further back in time. Finally, the common ancestor of all individuals (the point of convergence or coalescence) occurs at the date at the top of the figure. The actual genealogical tree of 134 individuals studied by Wilson is given in Figure 8.20 (the U shape is for compactness in representation; the diagram can be stretched out to resemble Figure 8.19). The most distant divergence was of six Africans (filled circles at the bottom right) from the rest of the population. Wilson therefore concluded that the earliest human split was between

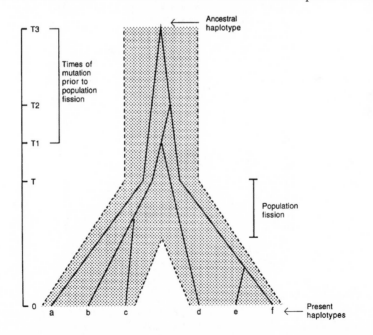

Figure 8.19. Diagrammatic representation of the coalescences of six individuals with different mitochondrial DNA into a single ancestral haplotype. The vertical axis represents time and the horizontal axis the degree of mtDNA differences.

Africans and non-Africans. This result agrees with that based on blood molecules: that modern humans originated in Africa. A multiregional development
of modern humans becomes even less likely. Figure 8.20 shows a horizontal
axis of sequence divergence that defines coalescence dates for other groups
(analogous to the vertical axis in Figure 8.19, although one represents time

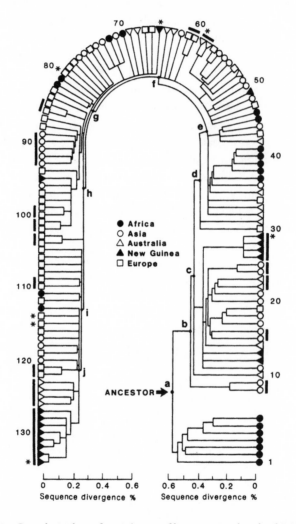

*Figure 8.20. Genealogical tree for 134 types of human mitochondrial DNA. The let-
ters (a–j) indicate points of fusion for populations, related to the degree of sequence
divergence noted on the horizontal axis. The shape of the point (circle, triangle, etc.)
indicates the geographical source. Black bars indicate clusters of mtDNA types specific
to a given geographic region. Asterisks indicate mtDNA types found in more than one
individual.*

and the other the proportion of mutations). The first common ancestor is labeled *a*. Ancestor *c* is the first to have no African descendants. From the percent divergence (0.43 percent, from the bar at the bottom), ancestor *c* lived 108,000 to 215,000 years ago (fractions of 140,000 and 290,000).

The Wilson study was the first of its kind and was subjected to considerable scrutiny and criticism. Although many difficulties were found, the conclusions have stood up very well. First, the rate of mutation was a particularly soft point. Subsequent studies have calculated the mutation rate from the divergence of orangutans from African apes or of humans from chimpanzees. The group of Satoshi Horai found that 3.89 to 7.00×10^{-8} change takes place per site per year. Second, the part of the mtDNA Wilson chose to study could have been atypical. He in fact looked at only about 9 percent of the mtDNA. The Horai study, however, examined every base pair of mtDNA in three individuals: one African, one Japanese, and one European. Their age of coalescence (Africans versus non-Africans) of 143,000 ± 18,000 years is just within Wilson's range. The small number of individuals is a limitation to this study. Horai also calculated that the age of coalescence of the European (Caucasian) and the Japanese (Northeast Asian) groups was 70,000 ± 13,000 years.

Acceptance of a coalescence of all human genes some 100,000 to 200,000 years ago still is not universal and has generated widespread misunderstanding. The work with mtDNA requires that all of this material in modern humans can be traced back to a single woman living at the point of coalescence. This mitochondrial Eve was not the only female ancestor at that time; nor was she possessed of some superior genetic property. The ultimate survival of her mtDNA was just a matter of chance. Best estimates are that 5,000 to 50,000 reproducing females were alive at that time. (The female population is estimated to be half of the coalescence time divided by the length of a generation, or 0.5[20,000/20] = 5000 for Wilson's calculation, which assumed a twenty-year generation.) Every time a female had no children or only male children, her mtDNA dropped out of the pool until, over many generations, everyone had the same mtDNA (aside from mutations). The same statistical process operates on surnames in societies in which the child inherits only the father's surname. The surname of each male without sons is lost, with the ultimate and inevitable consequence that eventually we all will have the same surname (unless the analogue of mutations occurs when people deviate from the normal naming process).

The male counterpart to mitochondrial Eve may be found by examining the coalescence of variation of the human Y chromosome, which is passed from fathers only to sons, just as mtDNA is passed from mothers only to daughters. Michael F. Hammer studied sixteen humans and four chimpanzees (as calibration, given the estimated distance in years to the human-chimpanzee split) and found a Y chromosome coalescence of 188,000 years and an effective male population of about five thousand. Thus the age of Y-chromosomal Adam agrees well with mitochondrial Eve. A second study based on only five humans, however, obtained a coalescence date of 37,000–49,000, although the smaller number of individuals renders this result less reliable.

Francisco J. Ayala argues for a much larger effective population of humans at the time of coalescence, about a hundred thousand. The population-genetics calculations give only the number of simultaneously reproducing individuals. Because many individuals would be too young, too old, or otherwise unable to reproduce, Ayala estimates that the above numbers represent only about two-ninths of the population, so that five thousand reproducing couples (ten thousand individuals) would be part of a population of 45,000.

Aside from constructing an overall genealogical framework for human history and providing estimates of population sizes, the analysis of modern DNA is contributing to archaeology in more specific ways. Rebecca Cann examined the mtDNA of thirty-five modern Hawaiians and found six individual lineages. Although this result indicates that all these Hawaiians are descended from six females, it does not mean that the original migratory canoes contained only six women. Some lineages may have become extinct, particularly after Western contact, and others may not have been sampled. In addition, many of the original Hawaiian women may have been matrilineally linked (sisters have the same mtDNA, as do first cousins descended matrilineally from a single grandmother, and so on). On the other hand, the study by Svante Pääbo and co-workers of the Nuu-Chah-Nulth tribe from Vancouver Island found considerable diversity, in terms of twenty-eight distinct lineages out of only sixty-three individuals, suggesting a large amount of populational mixing in this key migrational location. The distribution of the four aboriginal lineages (I–IV or A–D, now expanded to nine) has been interpreted by D. Wallace and others as supporting three ancient migrations into the New World (by the Amerind, the NaDene, and the Inuit or Eskimo groups), but D. Andrew Merriwether and co-workers see only one long migrational wave, and Antonio Torroni and co-workers prefer two migrations. Merriwether and Connie Kolman find the best genetic match between modern Native Americans and Mongolians rather than Siberians. The genealogy of the New World inhabitants still needs to be clarified.

The point of genetic coalescence (or divergence, depending on one's point of view) does not have to correspond to an actual period of migration, but it always must precede it. Thus an age of 100,000–200,000 years for mitochondrial Eve compares with the actual appearance of anatomically modern humans about 100,000 years ago. The genetic divergence of Southeast Asians from Australians and New Guineans occurred about 125,000 years ago, but the latter two land masses were inhabited starting as early as 55,000 years ago. Caucasoids separated from Northeast Asians and Amerinds about 80,000 years ago (as calculated by Cavalli-Sforza), but modern Europeans appeared as Cro-Magnons 35,000 to 45,000 years ago. On a more localized scale, Lapps separated from Caucasoids 26,000 years ago, but appeared in Lapland maybe 2,000 years ago; and Sardinians separated about 16,000 years ago, and settled there maybe 9,000 years ago. Thus genetic distances must be understood as referring to genetic events that always occurred prior to the archaeological events that are the more relevant focus for the reconstruction of human behavior. Applications of molecular biology to archaeology are still embryonic and will both enjoy expansion and suffer limitations as new data are obtained.

Epilogue

The human species has been set apart from other animals by our ability to make and utilize tools, leading Kenneth P. Oakley to coin the species-defining phrase "man the tool-maker." The production of tools, whether of wood, bone, or stone, originally involved only physical manipulations, sometimes of considerable delicacy and leading to aesthetically refined products. Numerous other species, however, can create simple tools, from birds that use twigs to uncover insects in bark to otters that use stones to break open shellfish. The human species, as the reader has seen in these chapters, became further distinguished from our fellow animals when we learned to manipulate the environment chemically. The controlled use of fire first provided protection and warmth, and later a means to preserve food. Eventually fire was used to rework the molecules of clay into pottery, of sand into glass, and of various minerals into metals and alloys. Grains were chemically transformed into bread and beer, skins and fibers into leather and linen. Resins from trees became adhesives, perfumes, and means for embalming, and bitumen became tar and pitch for sealing wooden surfaces of boats. All these discoveries and inventions were meeting the needs of our species thousands of years ago. Because humans improved tools by chemical modification (and fire certainly is a tool), perhaps another description of the human species is "the molecular transformer" or, more simply, "the chemical animal."

Glossary

absorption the process whereby certain wavelength ranges are removed as light passes through or reflects from a substance

acrylic a paint in which the binder is a material capable of forming a polymer in order to attach the pigment to the surface, normally with water as solvent

alcohol a class of organic compounds containing a hydroxy group

alkali metal the chemical elements found in the first column of the periodic table, most commonly lithium, sodium, and potassium

alkaline earth metal the chemical elements in the second column of the periodic table, most commonly magnesium, calcium, strontium, and barium

alkane a class of saturated organic molecules containing only carbon and hydrogen

allele one of various forms of a gene or of the characteristic it causes

alloy a solid mixture of two or more metals formed by fusion of the metals when molten

aluminosilicate a class of minerals composed of a silicate structure containing aluminum

amalgam an alloy of mercury with one or more metals

amalgamation reaction of a metal with mercury to form an amalgam, usually to extract the metal for later recovery through heating to drive off gaseous mercury

amber fossilized resin, formed from plant resins by the effects of temperature and pressure over geological time

amino acid a class of organic molecules containing both amino groups ($-NH_2$ or $-NH-$) and carboxyl groups ($-CO_2H$) that serve as the building blocks of proteins

anion a negatively charged chemical species

annealing heating a material to improve its properties, including the ability of stone to be fractured and of metal to be worked

asphalt naturally occurring or processed bitumen containing a mineral component such as calcite or silica, used for paving or as an adhesive

atom the smallest particle of an element

basalt a dense, igneous rock, usually gray to black in color

base pair a pair of side-chain bases on separate DNA strands that are connected by hydrogen bonds

bast fiber a fiber manufactured from the inner bark of plants, such as hemp

beer an alcoholic beverage made from malted grain, such as barley, that is flavored with hops and allowed to ferment

binder a substance used to attach a pigment to a substrate

bisque unglazed pottery that has been fired once

bitumen a naturally occurring tarry material, composed of hydrocarbons and lacking a mineral component

blast furnace a furnace in which a directed stream of air is used to increase the temperature and to enhance the efficiency of chemical reactions

bloom or bloomery iron impure iron produced without fusion by the direct reduction of iron oxides

bone ash the residue after bone is heated to a high enough temperature to remove organic components

bone china translucent china made with bone ash and the raw materials of porcelain

brass the alloy of 50–90 percent copper and 10–50 percent zinc, possessing a gold to red color

brick a usually rectangular object used for building or paving, made of any of several materials that include glass and pottery

calamine process a process for preparing brass from the reaction of copper with zinc carbonate (calamine)

cameo glass multilayered glass, in which the usually white outer layer is carved away to form a decoration in relief on a deeply colored background

carbohydrate a class of naturally occurring organic molecules that includes sugars as well as very large polymers (starch, cellulose) that are composed of sugar monomer units

carbonate a material that contains the carbonate group $CO_3{}^{2-}$

carbonyl group the chemical unit consisting of a carbon atom attached to an oxygen atom by two bonds, represented by $C=O$

cartouche an oblong shape in Egyptian art containing the hieroglyphic representation of a royal person's name

cased glass glass composed of two or more layers of different colors, often with carving to highlight inner layers

cast iron iron containing 2–5 percent carbon that is readily melted and poured into a mold but is brittle and unmalleable

cation a positively charged chemical species

celadon a ceramic with a gray-green glaze widely produced in China and Korea

cellulose a polysaccharide made up of units of the sugar glucose, constituting the chief component of plant cell walls and serving as the raw material for paper

cement a binding or structural material made from limestone and clay

cementation any process of heating a solid surrounded by a powder to effect their chemical combination; specifically, the methods to remove silver impurities from gold and to manufacture brass from the reaction of copper with zinc oxide

ceramic material produced by heating clay or other nonmetallic minerals

champlevé a type of enamel in which the raw material is applied and fired in cells that have been cut into the metal surface

china earthenware or porcelain tableware

china clay a term for kaolin, popular in England

china stone a term for porcelain stone, popular in England

cholesterol a twenty-seven-carbon alcohol present in animal fluids and cells

chromatography the procedure whereby mixtures of chemical compounds are separated into their pure forms by flowing over or through a liquid or solid material

cinnabar a pigment composed of mercury sulfide (HgS), also called vermilion

clay an earthy material composed of hydrated aluminosilicates and other minerals, formed by the natural weathering of rock; workable (plastic) when moist, it is able to harden with heating

cloisonné a type of enamel in which the raw material is applied and fired on a metal surface in raised cells separated by thin metal strips

cochineal a red dye produced from the body of the female insect *Dactylopius coccus*

cold-working hammering, bending, or otherwise processing metal at room temperature or at least below its temperature of crystallization, in order to harden and strengthen the material

collagen a protein present in connective tissue and bone, used as the raw material for glue

compound a chemical substance formed by the union of two or more different atoms in definite proportions

contact a term used to refer to the arrival of Europeans in the Americas

coprolite fossilized excrement

corrosion destruction by chemical action

creamware cream-colored, glazed earthenware

cross-linking the process of connecting one chain of atoms to another chain, thereby conveying strength and stability to a polymer

crucible steel steel produced by heating wrought iron with carbon in a crucible, used to make damascene

crystal glass   brilliantly clear, colorless glass of superior quality, usually containing high levels of lead

cultural diffusion   the slow movement of cultural elements from one area to another; contrasted to demic diffusion

cupellation   refinement of a precious metal such as silver by heating in a crucible, or cupel, with a stream of air to oxidize the baser metals, which are absorbed into the porous container

cut glass   glass with ornamentation cut into the surface, usually by an abrasive wheel

damascene   a steel product with wavy surface markings called damasks, made by a pattern-welding process and associated by Europeans with Damascus

decolorant   a material added to glass or another substance to counteract the coloring properties of existing components

demic diffusion   the slow movement of people from one area to another; contrasted to cultural diffusion

deoxyribonucleic acid (DNA)   the class of molecules that provides the basis of heredity

depletion gilding   chemical removal of an element from the surface of an object usually to achieve a desired surface color, as in the removal of copper from copper/gold mixtures to produce a gold surface

desaturation   loss of color intensity

diagenesis   reworking of the chemical constituents of a substance; the process whereby artifacts and animal remains are altered chemically by exposure to the environment during burial

dichroism   the property of exhibiting different colors to reflected and transmitted light

direct dye   a dye that may be applied to a substrate without chemical modification

double bond   a type of chemical connection between atoms consisting of two bonds, represented by two parallel connecting lines, as in $C=O$

drying oil   an oil used as a paint binder that dries and hardens, usually by polymerization, on exposure to air

ductility   the capacity of metal to be shaped through being pulled or drawn without structural rupture

dye   a coloring agent that is chemically bound to a substrate and permeates it fully

earthenware   slightly porous pottery generally fired in the range 900–1200°C to achieve a desired color, often covered with a glaze

electromagnetic spectrum   the entire wavelength range of light extending from gamma rays to radio waves, including visible light

electron spin resonance (ESR) a technique for studying unpaired electrons in a molecule; used to determine the age of an object

electrum a natural or synthetic alloy of gold containing about 20 percent or slightly more of silver, having a pale yellow color

element a substance that consists of atoms of only one kind

emulsion a stable dispersion of one liquid into another with which it is immiscible, in the form of very small droplets, such as fat in largely aqueous milk

enamel a glassy material attached to the surface of metal or other substances

enzyme a protein present in a living system to control and accelerate specific chemical reactions

fabric the body of a pottery object, not including surface additives

faience decorated, tin-glazed earthenware similar to majolica; an inappropriate term for glazed quartz from Egypt and elsewhere

famille rose/verte Chinese porcelain with a predominantly rose/green surface decoration

fat a water-insoluble class of naturally occurring molecules that are solid at normal temperatures and are composed of the chemical combination of the trialcohol glycerol and three fatty acids

fatty acid a class of organic molecules composed of a long hydrocarbon chain that terminates at one end with a carboxyl group ($-CO_2H$)

feldspar a group of aluminosilicate minerals containing sodium, potassium, or calcium

felt cloth made of wool or fur by the action of heat, moisture, chemicals, and especially pressure to create an interlocked mass

fermentation the biological conversion of sugar molecules to carbon dioxide and ethanol

filler temper

flint a hard form of quartz

fluorite the mineral calcium fluoride (CaF_2)

flux a substance used to promote the melting of another material

forging the working of usually hot metal through hammering in order to shape, purify, and strengthen the material

former the fundamental ingredient of glass that provides the silicate composition, such as quartz sand

fresco a painting produced by application of pigments to damp lime plaster

frit a partially fused, glassy substance used as raw material for making glass or glaze

fugitive likely to fade or disappear; said of a dye or pigment

gangue undesired minerals, such as silicates, found in ores

gas chromatography (GC) a form of chromatography in which the mixture to be separated is present as a gas

gilding   a process to give an object the appearance of gold, usually by overlaying it with a thin cover of gold

glass   any rigid material obtained by rapidly cooling a melt without crystallization, usually composed of silicates

glaze   a substance composed of quartz and other minerals, placed on the outside of pottery, and fired to produce a nonporous, glassy surface

glazed quartz   an object composed of a core of sintered quartz powder held together by a surface glaze; often given the misnomer *faience*

glucose · a widely occurring sugar with the formula $C_6H_{12}O_6$

glue   an adhesive made by cooking down collagen-containing materials such as hide and bone

gold filigree   gilding through the application of fine gold threads to the surface of an object

gold foil   thin gold sheet (at least 10^{-3} mm) used to cover an object and give it the appearance of gold

gold granulation   gilding through application of small beads of gold to the surface of an object

gold leaf   thin gold sheet (less than 10^{-3} mm) used to cover an object and give it the appearance of gold

gum   a class of water-soluble polysaccharides from plants that are gelatinous when moist but harden on drying

gum resin   resin that contains a large carbohydrate component

gunmetal   a copper alloy containing both tin and zinc, formerly used to fabricate cannon

haplotype   a set of genes inherited as a unit, or the visible characteristics produced by the genes

hectare   an area of 10,000 square meters or 2.47 acres

hematite   a red mineral composed of iron oxide with the formula Fe_2O_3, also called red ocher

horizon   a distinct layer of soil or other earthy material in a vertical section of land

hydrocarbon   a class of organic molecules containing only carbon and hydrogen

hydrogen bond   the weak interaction between a hydrogen atom and an atom such as oxygen or nitrogen that carries appreciable negative charge

hydroxy group   the chemical unit consisting of an oxygen atom attached to hydrogen, represented by OH

incunabula   books printed before A.D. 1501

index of refraction   a measure of the extent that light bends on going from a vacuum into a specific substance

infrared (IR) spectroscopy the study of light just beyond the red end of the visible region; used as a fingerprint of molecules

inglaze coloring applied in the glaze on a surface

ink a fluid used for writing or drawing, composed usually of water as solvent and a dye or pigment dispersed by means of a binder such as glue

inorganic derived from molecules based primarily on elements other than carbon

ion a charged chemical species, called an anion when the charge is negative or a cation when the charge is positive

isotopes forms of an element whose nuclei contain the same number of protons but different numbers of neutrons

isotopic fractionation the small increases in the proportion of one isotope over another of the same element during geological processes or chemical reactions that occur in living systems

jasper or jasperware colored stoneware with raised white decoration

jet a black gemstone related to coal

kaolin a fine, usually white clay used for the production of porcelain, composed principally of the mineral kaolinite

keratin a protein that is the main constituent of hair and nails

kermes a red dye produced from the body of the female insect *Kermes vermilio*

ketone a class of organic compounds containing a carbonyl group attached to two carbon atoms

kiln an enclosure for firing pottery with control of temperature and atmosphere

lacquer a durable coating formed by repetitively covering a surface with the sap from *Rhus vernicifera* and polishing

lake the combination of a dye and a mineral in the form of a powder or solid that may be used as a pigment

latex paint a paint in which the solvent is water

leather animal skin that has been processed for use

legume a plant belonging to the family Leguminosae that possesses root nodules capable of transforming atmospheric nitrogen into biologically useful molecules

leuco (or leuko) form the intermediate chemical form of a vat dye that has superior properties of solubility or attachment, often weak or lacking in color

lignin a large, irregular polymeric component of wood, composed of benzene rings linked by double bonds, oxygen atoms, and other groups

limestone a rock composed primarily of calcium carbonate and formed from the accumulated remains of marine shells and corals

lipid a large class of water-insoluble organic substances found in living cells, primarily including fats, oils, and waxes

litharge the monoxide of lead (PbO)

lost wax method a process of metal casting, whereby a wax model is covered with clay, the wax is melted and drained out to create a mold, and metal is poured into the empty mold to take the shape of the original wax model

lusterware glass with a surface composed of an often transparent metallic glaze that imparts an iridescent sheen

madder a red dye derived from the madder plant, *Rubia tinctorum;* also called Turkey red or alizarin, from one of its chemical components

majolica or maiolica earthenware covered with an opaque tin glaze, often with colored decorations applied before firing

malleability the capacity of metal to be shaped by hammering or rolling without structural rupture

malting the process of softening grain in water and allowing it to germinate to permit the breakdown of starch to simple sugars during brewing

marble a rock formed from limestone by metamorphosis

mass spectrometry (MS) a technique for measuring the molecular weight of molecules

metamorphosis the long-term effects of high temperature and pressure on rocks during burial in the earth

mica transparent silicate minerals that separate into thin sheets

midden an accumulation of debris or refuse from human activities, sometimes developing into a large mound

millefiori a glass object produced by cutting and fusing bundles of colored glass rods of various sizes and shapes

mitochondrion a cellular structure found outside the nucleus and used to produce energy

modifier a metal oxide added to silica to alter properties such as melting point without changing its fundamental character

molecular weight the total mass or weight of all the atoms in a molecule, expressed in units of the approximate weight of a hydrogen atom

molecule the smallest particle of a substance composed of two or more atoms

monomer a small compound that can combine with like monomers to form polymers

mordant a substance usually containing a metal that combines with a dye, rendering it insoluble and able to be attached to a substrate

mortar a pasty material that hardens and is used to bond two surfaces

mutation a change in hereditary material, including a physical change in a chromosome or a chemical change in the DNA

native metal naturally occurring pure metal, unbound to other elements

natron a mixture of minerals, but primarily sodium carbonate, used for embalming and glassmaking and for medical and detergent purposes

natural product any organic compound produced by a biological source

niello the process of packing incised designs on the surface of a silver, gold, or copper object with sulfides that are bonded to the surface through heating to form a black enamel-like decoration

nuclear magnetic resonance (NMR) spectroscopy a technique for identifying the structure of molecules by characterizing the carbons, hydrogens, and certain other constitutent atoms

obsidian natural glass formed from cooled lava

oil a class of organic molecules chemically identical to fats but existing as a liquid at normal temperatures

oil paint paint containing an organic-based binder such as linseed oil

opacifier a material added to glass or another substance to render it opaque

organic derived from molecules based primarily on the element carbon

oxidation an increase in the formal charge on an atom by removal of electrons; or, loosely, an increase in the number of attachments to oxygen or like atoms

oxidation state the formal charge on an atom, representing the number of electrons lost or gained in comparison with the original neutral atom

oxidizing conditions firing conditions in which there is an excess of atmospheric oxygen, that maintain high levels of oxygen attachments to metals

paint a mixture of pigment to provide color, binder to provide adherence, and solvent to provide fluidity during application to a surface

papyrus a sedge found in the Nile Valley and manufactured into a thin surface for writing

parcel-gilding a process for highlighting portions of a surface with a gold color

parchment the skin of sheep or goat prepared as a writing surface

parting a process for removing unwanted silver from gold

paste the material of the body or fabric of a pottery object, not including surface additives such as a slip or glaze

pattern welding a process for manufacturing iron or steel objects by repeated twisting, folding, and forging

petuntze porcelain stone

pewter the alloy of 75–95 percent tin and 5–25 percent lead, often used for domestic utensils and vessels

phosphate a material that contains the phosphate group PO_4^{3-}

phytolith a microscopic object made of silica (SiO_2), formed in plants and sometimes used as a means to determine the identity of plant remains in archaeological deposits

pigment a coloring agent that is applied to the surface of a substrate with the help of a binder

pitch the residue from heating natural organic materials, such as wood or peat, and driving off volatile components

plaster a pasty material that hardens on curing and is used to cover architectural surfaces

plastic capable of being molded

polyethylene glycol (PEG) an organic polymer used to preserve and strengthen woody artifacts

polymer a large molecule formed from repeating structural units or monomers

polymerase chain reaction (PCR) an enzymatic process for preparing large quantities of a segment of DNA from small amounts

polysaccharide a class of high-molecular-weight carbohydrates composed of numerous sugar monomer units

porcelain highly vitrified, nonporous, translucent, sonorous ceramic usually covered with a white glaze and fired above 1300°C

porcelain clay kaolin

porcelain stone a mineral containing quartz, mica, and feldspar used in production of porcelain

potash the mineral potassium carbonate (K_2CO_3), usually obtained from wood ash

pottery most generally, any ceramic material composed of clay and other constituents and hardened by heating

principal component the mathematical combination of several interdependent variables (such as elemental levels or haplotypes)

prospection in archaeology, exploration to locate sites of human activity or specific objects

protein a class of large, naturally occurring organic molecules, composed of amino acids, that provide biological structure and control biological reactions

provenance or provenience the original source of an artifact or of its raw materials

pseudomorph an originally biological material whose chemical constituents have been replaced entirely by minerals, while its appearance remains largely unchanged

pyrolysis chemical change caused by heating

pyrotechnology the creative application of fire in processing raw materials into useful products

quartzite a form of quartz (silicon dioxide) derived from sandstone by metamorphosis

racemization the process of conversion from the predominance of one non-superimposable mirror-image form of a molecule to an equal mixture of the two mirror-image forms

Raman spectroscopy a variety of infrared spectroscopy based on scattering rather than absorption of light

reducing conditions firing conditions with a deficit of atmospheric oxygen and heightened levels of carbon monoxide or hydrocarbons, resulting in reduced levels of oxygen attachments to metals

reduction a decrease in the formal charge on an atom by the addition of electrons; or, loosely, a decrease in the number of attachments to oxygen or like atoms

reflection the process whereby light returns from a surface or from within a substance with a simple change in direction

refraction the bending of light as it passes from one substance, such as air, into another, such as water

resin a sticky, water-insoluble plant secretion

roasting a chemical change brought about by heating an ore in air without melting, such as the conversion of sulfide ore to oxides

rosin the residue from heating pine resin

royal purple Tyrian purple

saccharide sugar

sandstone a sedimentary rock formed from quartz sand

sang-de-boeuf Chinese porcelain with a lustrous red glaze

saturated said of molecules lacking double and triple bonds

scanning electron microscope (SEM) an instrument for producing an enlarged image of a small object by means of a beam of electrons (in contrast to the use of light in the optical microscope)

scattering the process whereby light impinges on a substance and changes direction randomly

sediment largely inorganic material transported and deposited by the action of water, wind, or glacier

sericulture the production of silk from silkworms

sgraffito pottery with a slip surface in which decorations are cut to expose different colors, then usually fired with a glaze

silica a material composed of silicon dioxide (SiO_2), such as crystalline quartz

silicate a class of minerals containing chemical bonds between silicon and oxygen

sintering heating a material into a coherent mass without melting it

slag glassy waste product from smelting that includes silicates and iron oxides from the gangue as well as components of the flux

slip a suspension of clay in water, used to provide a thin layer (also called a slip) on pottery surface

smalt a blue pigment, discovered in the fifteenth century A.D., whose color derives from cobalt oxide

smelting the chemical processing, at high temperatures, of ores containing metals bound to nonmetallic elements in order to obtain a pure metal

soapstone   a soft stone having a soapy feel and composed of silicate minerals such as talc or chlorite

soda   the mineral sodium carbonate (Na_2CO_3)

soil   the surface layer of earth, composed of organic materials from decomposed organisms, and of inorganics, formed in place, although subject to weathering

solder   the alloy of 50–70 percent lead and 30–50 percent tin, used to join metals

stabilizer   a metal oxide added to silica to make the resulting glass more durable

starch   a complex, high-molecular-weight carbohydrate that is the storage form of sugars in plants

steatite   soapstone

steel   the alloy of iron with 0.2–2 percent carbon

stereochemistry   the arrangement of atoms in space within a molecule

stoneware   highly vitrified, nonporous, opaque ceramic fired in the range 1200–1300°C

stratigraphy   the arrangement of layers in an archaeological site caused by natural events and human occupation

substrate   the material on which a dye or pigment is placed

sugar   a water-soluble class of naturally occurring organic molecules made up of carbon, hydrogen, and oxygen, constituting a major food group

tanning   the process whereby hide is converted into leather

tar   volatile material condensed from vapors driven off from heated organics such as wood, coal, peat, or resin

tar pitch   the residue from heating tar to drive off volatiles

tell   a mound derived from the residue of successive human occupations

temper   inert, nonplastic material added to clay to minimize shrinkage and cracking during drying and firing, to modify plasticity, and to assist in vitrification

tempera   a paint that uses an oil-water emulsion as binder, the two substances being rendered miscible by an emulsifying agent such as egg yolk

tempering   in metallurgy, an annealing process to reduce brittleness of steel by heating and slow cooling

terpene   a class of hydrocarbons based on a five-carbon building block and found in natural oils and resins

terra-cotta   porous pottery made from clay and coarse tempers and fired usually around 900°C

textile   woven or knit cloth

thermoluminescence (TL)   the release of light that occurs when a substance is heated, used to determine the age of an object

tile   a thin flat or curved object made of ceramic, stone, or concrete and used for roofs, walls, or floors

transition metals   the chemical elements located centrally in the periodic table between scandium and zinc, yttrium and cadmium, and lanthanum and mercury, and possessing transitional properties

transmission   the process whereby light passes through a substance

triple bond   a type of chemical connection between atoms consisting of three bonds, represented by three parallel lines, as in C≡C

trophic level   the position of an animal on the food chain

tumbaga   an alloy of copper, gold, and possibly silver, discovered in the New World

tuyere   a pipe or nozzle through which air is delivered to a furnace or forge to raise its temperature

Tyrian purple   a purple dye produced from the shellfish *Murex* or *Purpuria*

ultraviolet (UV) spectroscopy   study of light just beyond the violet end of the visible region; used as a fingerprint of molecules

underglaze   decoration applied to the pottery surface before application of the glaze

unsaturated   said of molecules containing double or triple bonds

varnish   a clear, usually colorless liquid that is applied to a surface, which when dry provides protection or luster

vat dye   a water-insoluble dye that is altered chemically to a more workable substance, applied to a substrate, and then restored to its original chemical formulation

vehicle   a general term used for both the solvent and the binder in paint or ink, as the substance that carries the pigment

vellum   fine-grained parchment made from calfskin or lambskin

viticulture   the cultivation of grapes for wine-making

vitrification   formation of glassy substances with reduced porosity through melting and fusion of nonmetallic minerals at elevated temperatures

watercolor   paint composed of water as a solvent and a pigment suspended in the solvent by means of a binder, such as glue

wax   a chemically inhomogeneous class of organic materials composed of high-molecular-weight alcohols, fatty acids, and hydrocarbons, from both the plant and animal kingdoms, possessing a waxy character

wootz   crucible steel

wrought iron   the product of forging bloomery iron, containing less than 0.1 percent carbon but retaining 1–2 percent of silicate slag

X-ray diffraction   scattering of X-rays by atoms in a crystal, used to identify the chemical formula of minerals

Further Reading

Chapter 1: STONE

The Colossi of Memnon (Quartzite)

Bowman, H., F. H. Stross, F. Asaro, R. L. Hay, R. F. Heizer, and H. V. Michel. "The Northern Colossus of Memnon: New Slants." *Archaeometry,* vol. 26 (1984), pp. 218–29.

Marble

Craig, H., and V. Craig. "Greek Marbles: Determination of Provenance by Isotopic Analysis." *Science,* vol. 176 (1972), pp. 401–3.

Germann, J., G. Holzmann, and F. J. Winkler. "Determination of Marble Provenance: Limits of Isotopic Analysis." *Archaeometry,* vol. 22 (1980), pp. 99–106.

Herz, N. "Provenance Determination of Neolithic and Classical Mediterranean Marbles by Stable Isotopes." *Archaeometry,* vol. 34 (1992), pp. 185–94.

Herz, N., and D. B. Wenner. "Tracing the Origins of Marble." *Archaeology,* vol. 34, no. 5 (1981), pp. 14–21.

Flint

Bosch, P. W. "A Neolithic Flint Mine." *Scientific American,* vol. 240, no. 6 (1979), pp. 126–32.

Craddock, P. T., M. R. Cowell, M. N. Leese, and M. J. Hughes. "The Trace Element Composition of Polished Flint Axes as an Indicator of Source." *Archaeometry,* vol. 25, (1983), pp. 135–63.

Obsidian

Cann, J. R., J. E. Dixon, and C. Renfrew. "Obsidian Analysis and the Obsidian Trade." In *Science in Archaeology,* 2nd ed., edited by D. Brothwell and E. Higgs. New York: Praeger, 1970, pp. 578–91.

Gale, N. H. "Mediterranean Obsidian Source Characterisation by Strontium Isotope Analysis." *Archaeometry,* vol. 23 (1981), pp. 41–51.

Gordus, A. A., J. B. Griffin, and G. A. Wright. In *Science and Archaeology,* edited by R. H. Brill. Cambridge: MIT Press, 1971, pp. 222–34.

Healy, P. F., H. I. McKillop, and G. B. Walsh. "Analysis of Obsidian from Moho Cay, Belize: New Evidence on Classic Maya Trade Routes." *Science,* vol. 225 (1984), pp. 414–17.

Williams-Thorpe, O. "Obsidian in the Mediterranean and Near East: A Provenancing Success Story." *Archaeometry,* vol. 37 (1995), pp. 217–48.

Soapstone

Luckenback, A. H., C. G. Holland, and R. O. Allen. "Soapstone Artifacts: Tracing Prehistoric Trade Patterns in Virginia." *Science,* vol. 187 (1975), pp. 57–58.

Turquoise

Harbottle, G., and P. C. Weigard. "Turquoise in Pre-Columbian America." *Scientific American,* vol. 266, no. 2 (1992), pp. 78–85.

Basalt and Sandstone

Williams-Thorpe, O., and R. S. Thorpe. "Millstone Provenancing in Tracing the Route of a Fourth-Century B.C. Greek Merchant Ship." *Archaeometry,* vol. 32, (1990), pp. 115–37.

———. "Geochemistry, Sources and Transport of the Stonehenge Bluestones." In *New Developments in Archaeological Science,* edited by A.M. Pollard. *Proceedings of the British Academy,* vol. 77. Oxford, England: Oxford University Press, 1992, pp. 133–61.

Limestone and Calcite

Bello, M. A., and A. Martín. "Microchemical Characterization of Building Stone from Seville Cathedral, Spain." *Archaeometry,* vol. 34 (1992), pp. 21–29.

Hennig, G. J., W. Herr, E. Weber, and N. I. Xirotiris. "ESR-dating of the Fossil Hominid Cranium from Petralona Cave, Greece." *Nature,* vol. 292 (1981), pp. 533–36.

Herz, N., A. P. Grimanis, H. S. Rovinson, D. B. Wenner, and M. Vassilaki-Grimani. "Science *versus* Art History: The Cleveland Museum Head of Pan and the Miltiades Marathon Victory Monument." *Archaeometry,* vol. 31 (1989), pp. 161–68.

Holmes, L. L., G. Harbottle, and A. Blanc. "Compositional Characterization of French Limestone: A New Tool for Art Historians." *Archaeometry,* vol. 36 (1994), pp. 25–39.

Latham, A. G., and H. P. Schwarcz. "The Petralona Hominid Site: Uranium-Series Re-Analysis of 'Layer 10' Calcite and Associated Paleomagnetic Analyses." *Archaeometry,* vol. 34 (1992), pp. 135–40.

Rock Varnish

Dorn, R. I. "Rock Varnish." *American Scientist,* vol. 79 (1991), pp. 542–53.

Stone Conservation

Fleming, S. "Crumbling Facades: A Story of Pollution." *Archaeology,* vol. 35, no. 5 (1982), pp. 72–73.

Freund, E. C. "Saving the Monuments of the Athenian Acropolis." In *Ancient Technologies and Archaeological Materials,* edited by S. U. Wisseman and W. S. Williams. Langhorne, Pa.: Gordon and Breach, 1994, pp. 199–215.

Lal Gauri, K. "The Preservation of Stone." *Scientific American,* vol. 238, no. 6 (1978), pp. 126–36.

Synthetic Stone

Kingery, W. D., P. B. Vandiver, and M. Prickett. "The Beginnings of Pyrotechnology, Part II: Production and Use of Lime and Gypsum Plaster in the Pre-Pottery Neolithic Near East." *Journal of Field Archaeology,* vol. 15 (1988), pp. 219–44.

Jedrzejewska, H. "New Methods in the Investigation of Ancient Mortars." In *Archaeological Chemistry,* edited by M. Levey. Philadelphia: University of Pennsylvania Press, 1967, pp. 147–66.

Rollefson, G. O. "The Uses of Plaster at Neolithic 'Ain Ghazhal, Jordan." *Archaeo-materials,* vol. 4 (1990), pp. 33–43.

Vendrell-Saz, M., S. Alarcón, J. Molera, and M. García-Vallés. "Dating Ancient Lime Mortars by Geochemical and Mineralogical Analysis." *Archaeometry,* vol. 38 (1996), pp. 143–49.

Chapter 2: SOIL

General

Cornwall, I. W. "Soil, Stratification, and Environment." In *Science and Archaeology,* edited by D. Brothwell and E. Higgs, New York: Praeger, 1970, pp. 124–34.

Limbrey, S. *Soil Science and Archaeology,* London: Academic Press, 1975.

Prospecting with Phosphorus

Bethell, P., and I. Máté. "The Use of Soil Phosphate Analysis in Archaeology: A Critique." In *Scientific Analysis in Archaeology,* edited by J. Henderson. Oxford: Oxford University Committee for Archaeology, 1989, pp. 1–29.

Eidt, R. C. "Detection and Examination of Anthrosols by Phosphate Analysis." *Science,* vol. 197 (1977), pp. 1327–33.

Proudfoot, B. "The Analysis and Interpretation of Soil Phosphorus in Archaeological Contexts." In *Geoarchaeology: Earth Science and the Past,* edited by D. A. Davidson and M. L. Shackley. London: Duckworth, 1976, pp. 93–113.

Vertical Surveys

Davidson, D. A. "Particle Size and Phosphate Analysis—Evidence for the Evolution of a Tell." *Archaeometry,* vol. 15 (1973), pp. 143–52.

Sánchez, A., M. L. Cañabate, and R. Lizcano. "Phosphorus Analysis at Archaeological Sites: An Optimization of the Method and Interpretation of the Results." *Archaeometry,* vol. 38 (1996), pp. 151–64.

Prospecting with Other Elements

Woods, W. I. "Soil Chemistry Investigations in Illinois Archaeology: Two Example Studies." In *Archaeological Chemistry III,* edited by J. B. Lambert, Washington, D.C.: American Chemical Society, 1984, pp. 67–77.

Analysis for Specific Human Activities

Conway, J. S. "An Investigation of Soil Phosphate Distribution within Occupational Deposits from a Romano-British Hut Group." *Journal of Archaeological Science,* vol. 10 (1983), pp. 117–28.

Cook, S. F., and R. F. Heizer. *Studies on the Chemical Analysis of Archaeological Sites.* Berkeley, Calif.: University of California Press, 1965, pp. 4–8.

Craddock, P. T., D. Gurney, F. Pryor, and M. J. Hughes. "The Application of Phosphate Analysis to the Location and Interpretation of Archaeological Sites." *The Archaeological Journal,* vol. 142 (1985), pp. 361–76.

Konrad, V. A., R. Bonnichsen, and V. Clay. "Soil Chemistry Identification of Ten Thousand Years of Prehistoric Human Activity at the Munsungen Lake Thoroughfare, Maine." *Journal of Archaeological Science,* vol. 10 (1983), pp. 13–28.

Organic Residues in Soil

Bethell, P. H., L. J. Goad, R. P. Evershed, and J. Ottaway. "The Study of Molecular Markers of Human Activity: The Use of Coprostanol in the Soil as an Indicator of Human Faecal Material." *Journal of Archaeological Science,* vol. 21 (1994), pp. 619–32.

Skeletal Silhouettes

Biek, L. "Soil Silhouettes." In *Science in Archaeology,* edited by D. Brothwell and E. Higgs, New York: Praeger, 1970, pp. 118–23.

Keeley, H. C. M., G. E. Hudson, and J. Evans. "Trace Element Contents of Human Bones in Various States of Preservation 1. The Soil Silhouette." *Journal of Archaeological Science,* vol. 4 (1977), pp. 19–24.

Lambert, J. B., S. V. Simpson, J. E. Buikstra, and D. K. Charles. "Analysis of Soil Associated with Woodland Burials." In *Archaeological Chemistry III,* edited by J. B. Lambert, Washington, D.C.: American Chemical Society, 1984, pp. 97–113.

Soil Stratigraphy

Allen, R. O., H. Hamroush, and M. A. Hoffman. "Archaeological Implications of Differences in the Composition of Nile Sediments." In *Archaeological Chemistry IV*, edited by R. O. Allen. Washington, D.C.: American Chemical Society, 1989, pp. 33–56.

Stimmell, C. A., R. G. V. Hancock, and A. M. Davis. "Chemical Analysis of Archaeological Soils from Yagi Site, Japan." In *Archaeological Chemistry III*, edited by J. B. Lambert, Washington, D.C.: American Chemical Society, 1984, pp. 79–96.

Chapter 3: POTTERY

General

Cooper, E. *A History of Pottery*, London: Longman, 1972.

Cottier-Angeli, F. *Ceramics*. New York: Van Nostrand Reinhold, 1974.

Neff, H., ed. *Chemical Characterization of Ceramic Pastes in Archaeology*. Monographs in World Archaeology No. 7. Madison, Wis.: Prehistory Press, 1992.

Nelson, G. C. *Ceramics*, New York: Holt, Rinehart and Winston, 1960.

Olin, J. S., and A. P. Franklin, eds. *Archaeological Ceramics*. Washington, D.C.: Smithsonian Institution Press, 1982.

Orton, C., P. Tyers, and A. Vince. *Pottery in Archaeology*. Cambridge, England: Cambridge University Press, 1993.

Rice, P. M., ed. *Pots and Potters*. University of California, Los Angeles: Institute of Archaeology, 1984.

Rice, P. M. *Pottery Analysis: A Sourcebook*. Chicago: University of Chicago Press, 1987.

Sinopoli, C. M. *Approaches to Archaeological Ceramics*. New York: Plenum Press, 1991.

Japanese Ceramics

Ikawa-Smith, F. "Current Issues in Japanese Archaeology." *American Scientist*, vol. 68 (1980), pp. 134–45.

Egyptian Ceramics

Kaplan, M. F., G. Harbottle, and E. V. Sayre. "Multi-Disciplinary Analysis of Tell el Yahudiyeh Ware." *Archaeometry*, vol. 24 (1982), pp. 127–42.

Mycenaean Ceramics

Catling, H. W., and A. Millett. "A Study of the Inscribed Stirrup-Jars from Thebes." *Archaeometry*, vol. 8 (1965), pp. 3–85.

Catling, H. W., E. E. Richards, and A. E. Blin-Stoyle. "Correlations between Composition and Provenance of Mycenaean and Minoan Potter." *Annual of the British School of Archaeology at Athens*, vol. 58 (1963), pp. 94–115.

Lambert, J. B., C. D. McLaughlin, and A. Leonard, Jr. "X-Ray Photoelectron Spectroscopic Analysis of the Mycenaean Pottery from Megiddo." *Archaeometry*, vol. 20 (1978), pp. 107–22.

Classic Greek Ceramics

Cook, R. M. *Greek Painted Pottery*, 2nd ed. New York: Methuen, 1972.
Noble, J. V. *Techniques of Attic Painted Pottery*, London: Thames and Hudson, 1988.
Sparkes, B. A. *Greek Pottery: An Introduction.* Manchester: Manchester University Press, 1991.

Majolica and Related Materials

Fleming, S. "Josiah Wedgwood: A Potter of Fashion." *Archaeology*, vol. 39, no. 3 (1986), pp. 70, 71, 81.
Godden, G. A. *British Pottery: An Illustrated Guide.* London: Barrie and Jenkins, 1974.
Kleinmann, B. "History and Development of Early Islamic Pottery Glazes." In *Proceedings of the 24th International Archaeometry Symposium,* edited by J. S. Olin and M. J. Blackman. Washington, D.C.: Smithsonian Institution Press, 1966, pp. 73–76.
Olin, J. S., and M. J. Blackman. "Compositional Classification of Mexican Majolica Ceramics of the Spanish Colonial Period." In *Archaeological Chemistry IV,* edited by R. O. Allen. Washington, D.C.: American Chemical Society, 1989, pp. 87–112.
Reilly, R. *Wedgwood Jasper.* New York: Thames and Hudson, 1994.
Scovizzi, G. *Maiolica, Delft, and Faïence.* London: Hamlyn, 1966.
Whitehouse, D. "The Origins of Italian Maiolica." *Archaeology*, vol. 31, no. 2 (1978), pp. 42–49.

Porcelain

Berges, R. *From Gold to Porcelain.* New York: Thomas Yoseloff, 1963.
Fisher, S. *English Ceramics.* New York: Hawthorn Books, 1965.
Guo, Y., "Raw Materials for Making Porcelain and the Characteristics of Porcelain Wares in North and South China in Ancient Times." *Archaeometry*, vol. 29 (1987), pp. 3–19.
Neave-Hill, W. B. R. *Chinese Ceramics.* Edinburgh: John Bartholemew, 1975.
Satō, M. *Chinese Ceramics.* New York: Weatherhill, 1978.
Tite, M. S., and M. Bimson. "A Technological Study of English Porcelains." *Archaeometry*, vol. 33 (1991), pp. 3–27.

Chapter 4: COLOR

General

McClaren, K. *The Colour Science of Dyes and Pigments.* Bristol, England: Adam Hilger, 1983.

Travis, A. S. *The Rainbow Makers*. Bethlehem, Pa.: Lehigh University Press, 1993.
Zollinger, H. *Color Chemistry,* 2nd ed. Weinheim, GermaN.Y.: VCH, 1991.

Egyptian, Mesopotamian, and Early Greek Pigments

Chase, W. T. "Egyptian Blue as a Pigment and Ceramic Material." In *Science and Archaeology,* edited by R. H. Brill. Cambridge: MIT Press, 1971, pp. 80–90.

Hassan, A. A., and F. A. Hassan. "Source of Galena in Predynastic Egypt at Nagada." *Archaeometry,* vol. 23 (1981), pp. 77–82.

Kaczmarczyk, A. "The Source of Cobalt in Ancient Egyptian Pigments." In *Proceedings of the 24th International Archaeometry Symposium,* edited by J. S. Olin and M. J. Blackman. Washington, D.C.: Smithsonian Institution Press, 1986, pp. 369–76.

Nir-el, Y., and M. Broshi. "The Red Ink of the Dead Sea Scrolls." *Archaeometry,* vol. 38 (1996), pp. 97–102.

Noll, W., R. Holm, and L. Born. "Painting of Ancient Ceramics." *Angewandte Chemie, International Edition (English),* vol. 14 (1975), pp. 602–13.

Riederer, H. "Recently Identified Egyptian Pigments." *Archaeometry,* vol. 16 (1974), pp. 102–9.

Tite, M. S., M. Bimson, and M. R. Cowell. "Technological Examination of Egyptian Blue." In *Archaeological Chemistry III,* edited by J. B. Lambert. Washington, D.C.: American Chemical Society, 1984, pp. 215–42.

Ancient Dyes from Europe and Southwest Asia

Foster, G. V., and P. J. Moran. "Plants, Paints, and Pottery: Identification of Madder Pigment on Cypriot Ceramicware." In *Archaeometry,* edited by Y. Maniatis, Amsterdam: Elsevier, 1989, pp. 183–89.

Green, L. R., and V. Daniels. "Shades of the Past." *Chemistry in Britain,* (1995), pp. 613–16.

Hoffmann, R. "Blue as the Sea." *American Scientist,* vol. 78 (1990), pp. 308–9.

McGovern, P. E., and R. H. Michael. "Royal Purple Dye: Tracing Chemical Origins of the Industry." *Analytical Chemistry,* vol. 57 (1985), pp. 1514A-22A.

———. "Royal Purple Dye: The Chemical Reconstruction of the Ancient Mediterranean Industry." *Accounts of Chemical Research,* vol. 23 (1990), pp. 152–58.

Pre-Columbian Dyes and Pigments of the New World

Arnold, D. E., and B. F. Bohor. "Attapulgite and Maya Blue." *Archaeology,* vol. 38, no. 1 (1975), pp. 23–29.

Fleming, S. "The Tale of the Cochineal: Insect Farming in the New World." *Archaeology,* vol. 36, no. 5 (1983), pp. 68–69, 79.

Jakes, K. A., J. E. Katon, and P. A. Martoglio. "Identification of Dyes and Characterization of Fibers by Infrared and Visible Microspectroscopy: Application to Paracas Textiles." In *Archaeometry '90,* edited by E. Pernicka and G. A. Wagner. Basel: Birkhäuser Verlag, 1991, pp. 305–15.

José-Yacamán, M., L. Rendón, J. Arenas, and M. C. Serra Puche. "Maya Blue: An Ancient Nanostructured Material." *Science,* vol. 273 (1996), pp. 223–25.

Saltzman, M., "Identifying Dyes in Textiles." *American Scientist,* vol. 80 (1992), pp. 474–81.

Asian Materials

Fleming, S. " *Ukiyo-e* Painting: An Art Tradition Under Stress." *Archaeology,* vol. 38, no. 6 (1985), pp. 60–61, 75.

Kleinmann, B. "Cobalt-Pigments in the Early Islamic Blue Glazes and the Reconstruction of the Way of Their Manufacture." In *Archaeometry '90,* edited by E. Pernicka and G. A. Wagner. Basel: Birkhäuser Verlag, 1991, pp. 327–36.

Winter, J. "Preliminary Investigations on Chinese Ink in Far Eastern Paintings." In *Archaeological Chemistry,* edited by C. W. Beck. Washington, D.C.: American Chemical Society, 1974, pp. 207–25.

Medieval Dyes and Pigments

Cahill, T. A., B. H. Kusko, R. A. Eldred, and R. N. Schwab. "Gutenberg's Inks and Papers: Non-Destructive Compositional Analyses by Proton Milliprobe." *Archaeometry,* vol. 26 (1984), pp. 3–14.

Fleming, S. "Pigments in History: The Growth of the Traditional Palette." *Archaeology,* vol. 37, no. 4 (1984), pp. 68–69.

Harley, R. D. *Artists' Pigments: c. 1600–1835.* London: Butterworths, 1970.

Pellicori, S., and M. S. Evans. "The Shroud of Turin through the Microscope." *Archaeology,* vol. 34, no. 1 (1981), pp. 34–43.

Chapter 5: GLASS

General

Brill, R. H. "Ancient Glass." *Scientific American,* vol. 209, no. 5 (1963), pp. 120– 26.

Frank, S. *Glass and Archaeology.* London: Academic Press, 1982.

Neuburg, F. *Ancient Glass.* London: Barrie & Rockliff, 1962.

Sayre, E. V., and R. W. Smith. "Compositional Categories of Ancient Glass." *Science,* vol. 133 (1961), pp. 1824–26.

Wertime, T. A. "Pyrotechnology: Man's First Industrial Use of Fire." *American Scientist,* vol. 61 (1973), pp. 670–82.

Glazed Quartz

Aspinall, A., S. E. Warren, J. G. Crummett, and R. G. Newton. "Neutron Activation Analysis of Faience Beads." *Archaeometry,* vol. 14 (1972), pp. 27–40.

Kaczmarcyzk, A., and R. E. M. Hedges. *Ancient Egyptian Faience: An Analytical Survey of Egyptian Faience from Predynastic to Roman Times.* Warminster: Aris & Phillips, 1983.

Kiefer, C., and A. Allibert. "Pharaonic Blue Ceramics: The Process of Self-Glazing." *Archaeology*, vol. 24, no. 2 (1971), pp. 107–17.

Tite, M. S. "Characterisation of Early Vitreous Materials." *Archaeometry*, vol. 29 (1987), pp. 21–34.

Vandiver, P. "Egyptian Faience Technology." In *Archaeological Ceramics*, edited by A. P. Franklin and J. Olin. Washington, D.C.: Smithsonian Institution Press, 1982.

Egyptian, Mesopotamian, and Syrian Glass

Oppenheim, A. L., R. H. Brill, D. Barag, and A. von Saldern. *Glass and Glassmaking in Ancient Mesopotamia*. Corning, N.Y.: Corning Museum of Glass Press, 1970.

Roman Glass

Barber, D. J., and I. C. Freestone. "An Investigation of the Origin of the Colour on the Lycurgus Cup by Analytical Transmission Electron Microscopy." *Archaeometry*, vol. 32 (1990), pp. 33–45.

Goldstein, S. M. "2,000 Years of Cameo Glass at the Corning Museum." *Archaeology*, vol. 35, no. 4 (1982), pp. 54–57.

Grose, D. "The Formation of the Roman Glass Industry." *Archaeology*, vol. 36, no. 4 (1983), pp. 38–45.

Harden, D. B., J. Hellenkemper, K. Painter, and D. Whitehorse. *Glass of the Caesars*. Milan: Olivetti, 1987.

Asian Glass

Brill, R. H. "Chemical Analysis of Some Early Indian Glasses." In *Archaeometry of Glass. Proceedings of the Archaeometry Session of the XIV International Congress on Glass 1986 New Delhi India*, edited by H. C. Bhardwaj. Calcutta: Indian Ceramic Society, 1987, pp. 247–92.

———. "Scientific Investigations of Ancient Asian Glass." In *Nara Symposium '91, Report, UNESCO Maritime Route of Silk Roads, Nara*. 1993, pp. 70–79.

———. "Glass and Glassmaking in Ancient China, and Some Other Things from Other Places." *The Glass Art Society Journal*, 1993, pp. 163–176.

Brill R. H., and J. H. Martin, eds. *Scientific Research in Early Chinese Glass*. Corning, N.Y.: Corning Museum of Glass, 1991.

Henderson, J., M. Tregear, and N. Wood. "The Technology of Sixteenth- and Seventeenth-Century Chinese *Cloisonné* enamels." *Archaeometry*, vol. 31 (1989), pp. 133–46.

Yamasaki, K. "Technical Studies on the Ancient Art Objects in Japan, with Special Reference to the Treasures Preserved in the Shōsō-in." In *Application of Science in Examination of Works of Art*, edited by P. T. Rathbone and W. J. Young. Boston: Museum of Fine Arts, 1967, pp. 114–25.

European Glass

Brill, R. H. "Chemical Analyses of Some Glasses from Frattesina." *Journal of Glass Studies,* vol. 34, page 11–22.

Brun, N., and M. Pernot. "The Opaque Red Glass of Celtic Enamels from Continental Europe." *Archaeometry,* vol. 34 (1992), pp. 235–52.

Henderson, J. "Electron Probe Microanalysis of Mixed-Alkali Glasses." *Archaeometry,* vol. 30 (1988), pp. 77–91.

———. "The Scientific Analysis of Ancient Glass and its Archaeological Interpretation." In *Scientific Analysis in Archaeology,* edited by J. Henderson. Oxford: Oxford University Press, 1989, pp. 30–62.

Olin, J. S., and E. V. Sayre. "Neutron Activation Analytical Survey of Some Intact Medieval Glass Panels and Related Specimens." In *Archaeological Chemistry,* edited by C. W. Beck. Washington, D.C.: American Chemical Society, 1974, pp. 100–23.

Tennent, N. H., P. McKenna, K. K. N. Lo, G. McLean, and J. M. Ottaway. "Major, Minor, and Trace Element Analysis of Medieval Stained Glass by Flame Atomic Absorption Spectroscopy." In *Archaeological Chemistry III,* edited by J. B. Lambert, Washington, D.C.: American Chemical Society, 1984, pp. 133–50.

Chapter 6: ORGANICS

Fire

Brain, C. K., and A. Sillen. "Evidence from the Swartkrans Cave for the Earliest Use of Fire." *Nature,* vol. 336 (1988), pp. 464–66.

Food

Biers, T. W., and P. E. McGovern, eds. *Organic Contents of Ancient Vessels. MAS CA Research Papers in Science and Archaeology,* vol. 7, Philadelphia: University Museum, 1990.

Evershed, R. P., C. Heron, S. Charters, and L. J. Goad. "The Survival of Food Residues: New Methods of Analysis, Interpretation and Application." In *New Developments in Archaeological Science,* edited by A. M. Pollard. Oxford: Oxford University Press, 1992, pp. 187–208.

Hastorf, C. A., and M. J. DeNiro. "Reconstruction of Prehistoric Plant Production and Cooking Practices by a New Isotopic Method." *Nature,* vol. 315 (1985), pp. 489–91.

Heiser, C. B., Jr. *Seed to Civilization,* 3rd ed. Cambridge: Harvard University Press, 1990.

Oudemans, T. F. M., and J. J. Boon. "Molecular Archaeology: Analysis of Charred (Food) Remains from Prehistoric Pottery by Pyrolysis-Gas Chromatography/Mass Spectrometry." *Journal of Analytical Applied Pyrolysis,* vol. 20 (1991), pp. 197–227.

Beer and Wine

Brown, S. C. "Beers and Wines of Old New England." *American Scientist,* vol. 66, (1978), pp. 460–67.

Katz, S. L., and F. Maytag. "Brewing an Ancient Beer." *Archaeology,* vol. 44, no. 4 (1991), pp. 24–33.

McGovern, P. E., D. L. Gusker, L. J. Exner, and M. M. Voigt. "Neolithic Resinated Wine." *Nature,* vol. 381 (1996), pp. 480–81.

Sams, G. K. "Beer in the City of Midas." *Archaeology,* vol. 30, no. 2 (1977), pp. 108–15.

Samuel, D. "Investigation of Ancient Egyptian Baking and Brewing Methods by Correlative Microscopy." *Science,* vol. 273 (1996), pp. 488–90.

Unwin, T. *Wine and the Vine,* London: Routledge, 1991.

Webb, A. D. "The Science of Making Wine." *American Scientist,* vol. 72 (1984), pp. 360–67.

Wiseman, J. "To Your Health!" *Archaeology,* vol. 49, no. 2 (1996), pp. 23–26.

Fibers and Textiles

Barber, E. J. W. *Prehistoric Textiles.* Princeton, N.J.: Princeton University Press, 1991.

Betts, A., K. van der Borg, A. de Jong, C. McClintock, and M. van Strydonck. "Early Cotton in North Arabia." *Journal of Archaeological Science,* vol. 21 (1994), pp. 489–99.

Geijer, A. *A History of Textile Art.* Stockholm: Paold Research Fund, 1979.

Jakes, K. A., and L. R. Sibley. "An Examination of the Phenomenon of Textile Fabric Pseudomorphism." In *Archaeological Chemistry III,* edited by J. B. Lambert. Washington, D.C.: American Chemical Society, 1984, pp. 403–24.

Proefke, M. L., K. L. Rinehart, M. Raheel, S. H. Ambrose, and S. U. Wisseman. "Probing the Mysteries of Ancient Egypt: Chemical Analysis of a Roman Period Egyptian Mummy." *Analytical Chemistry,* vol. 64 (1992), pp. 105A–11A.

Raheel, M. "History, Identification, and Characterization of Old World Fibers and Dyes." In *Ancient Technologies and Archaeological Materials,* edited by S. U. Wisseman and W. S. Williams. Langhorne, Pa.: Gordon and Breach Science Publishers, 1994, pp. 121–53.

Skins

Reed, R. *Ancient Skins, Parchments and Leathers.* London: Seminar Press, 1972.

Ryder, M. L., "Remains Derived from Skin." In *Science in Archaeology,* edited by D. Brothwell and E. Higgs. New York: Praeger, 1970, pp. 539–54.

Skelton, R. A., T. E. Marston, and G. D. Printer. *The Vinland Map and the Tartar Relation,* 2nd ed. New Haven: Yale University Press, 1995.

Bone and Ivory

Baer, N. S., T. Jochsberger, and N. Indictor. "Chemical Investigations of Ancient Near Eastern Archaeological Ivory Artifacts. Fluorine and Nitrogen Composition." In

Archaeological Chemistry II, edited by G. F. Carter. Washington, D.C.: American Chemical Society, 1978, pp. 137–49.

Chaplin, R. E. *The Study of Animal Bones from Archaeological Sites.* London: Seminar Press, 1971.

Van der Merwe, N. J., L. J. Lee-Thorpe, J. F. Thackeray, A. Hall-Martin, F. J. Kruger, H. Coetzee, R. H. V. Bell, and M. Lideque. "Source-Area Determination of Elephant Ivory by Isotopic Analysis." *Nature,* vol. 346 (1990), pp. 744–46.

Wood and Paper

Barbour, R. J. "Treatments for Waterlogged and Dry Archaeological Wood." In *Archaeological Wood Properties, Chemistry, and Preservation,* edited by R. M. Rowell and R. J. Barbour. Washington, D.C.: American Chemical Society, 1990, pp. 177–92.

Håfors, B. "The Role of the *Wasa* in the Development of the Polyethylene Glycol Preservation Method." In *Archaeological Wood Properties,* edited by R. M. Rowell and R. J. Barbour. Washington, D.C.: American Chemical Society, 1990, pp. 195–216.

Hedges, J. I. "The Chemistry of Archaeological Wood." In *Archaeological Wood Properties,* edited by R. M. Rowell and R. J. Barbour, Washington, D.C.: American Chemical Society, 1990, pp. 111–40.

Kaye, B. "Conservation of Waterlogged Archaeological Wood." *Chemical Society Reviews,* vol. 24 (1995), pp. 35–43.

Lewis, N. "Papyrus and Ancient Writing: The First Hundred Years of Papyrology." *Archaeology,* vol. 36, no. 4 (1990), pp. 31–37.

Natural Products

Beck, C. W. "Spectroscopic Investigations of Amber." *Applied Spectroscopy Reviews,* vol. 22 (1986), pp. 57–110.

Cassar, M., G. V. Robins, R. A. Fletton, and A. Alstin. "Organic Components in Historical Non-metallic Seals Identified Using ^{13}C-NMR spectroscopy." *Nature,* vol. 303 (1983), pp. 238–39.

Charters, S., R. P. Evershed, L. J. Goad, C. Heron, and P. Blinkhorn. "Identification of an Adhesive Used to Repair a Roman Jar." *Archaeometry,* vol. 35 (1993), pp. 91–101.

Evans, K., and C. Heron. "Glue, Disinfectant and Chewing Gum: Natural Products Chemistry in Archaeology." *Chemistry & Industry,* (1993), pp. 446–49.

Lambert, J. B., C. W. Beck, and J. S. Frye. "Analysis of European Amber by Carbon-13 Nuclear Magnetic Resonance Spectroscopy." *Archaeometry,* vol. 30 (1988), pp. 248–63.

Marschner, R. F., and H. T. Wright. "Asphalts from Middle Eastern Archaeological Sites." In *Archaeological Chemistry II,* edited by G. F. Carter. Washington, D.C.: American Chemical Society, 1978, pp. 150–71.

Mills, J. S., and R. White. "Natural Resins of Art and Archaeology. Their Sources, Chemistry, and Identification." *Studies in Conservation,* vol. 22 (1977), pp. 12–31.

———. *The Organic Chemistry of Museum Objects,* 2nd ed. Oxford: Oxford University Press, 1994.

Muller, H. *Jet.* London: Butterworths, 1987.

Robinson, N., R. P. Evershed, W. J. Higgs, K. Jerman, and G. Eglinton. "Proof of a Pine Wood Origin for Pitch from Tudor (*Mary Rose*), and Etruscan Shipwrecks: Application of Analytical Organic Chemistry in Archaeology." *Analyst,* vol. 112 (1987), pp. 637–44.

Snyder, D. M. "An Overview of Oriental Lacquer." *Journal of Chemical Education,* vol. 66 (1989), pp. 977–80.

Twilley, J. W. "The Analysis of Exudate Plant Gums in Their Artistic Applications: An Interim Report." In *Archaeological Chemistry III,* edited by J. B. Lambert. Washington, D.C.: American Chemical Society, 1984, pp. 357–94.

White, R. "The Application of Gas-Chromatography to the Identification of Waxes," *Studies in Conservation,* vol. 23 (1978), pp. 57–68.

Chapter 7: METALS

Native Copper

Hancock, R. G. V., L. A. Pavlish, R. M. Farquhar, R. Salloum, W. A. Fox, and G. C. Wilson. "Distinguishing European Trade Copper and Northeastern North American Native Copper." *Archaeometry,* vol. 33 (1991), pp. 69–86.

Rapp, G., Jr., J. Allert, and E. Henrickson. "Trace Element Discrimination of Discrete Sources of Native Copper." In *Archaeological Chemistry III,* edited by J. B. Lambert. Washington, D.C.: American Chemical Society, 1984, pp. 273–93.

Smith, C. S. "The Interpretation of Microstructures of Metallic Artifacts." In *Application of Science in Examination of Works of Art,* edited by P. T. Rathbone and W. J. Young. Boston: Museum of Fine Arts, 1967, pp. 20–52.

Wertime, T. A. "The Beginnings of Metallurgy: A New Look." *Science,* vol. 182 (1973), pp. 875–87.

Copper Smelting

Glumac, P. D., and J. A. Todd. "Eneolithic Copper Smelting Slags from the Middle Danube Basin." In *Archaeometry '90,* edited by E. Pernicka and G. A. Wagner. Basel: Birkhäuser Verlag, 1991, pp. 155–64.

Goffer, Z. *Archaeological Chemistry.* New York: Wiley, 1980, chap. 11.

Hong, S., J.-P. Candelone, C. C. Patterson, and C. F. Boutron. "History of Copper Smelting Pollution During Roman and Medieval Times Recorded in Greenland Ice." *Science,* vol. 272 (1996), pp. 246–49.

Lechtman, H. "The Central Andes: Metallurgy without Iron." In *The Coming of the Age of Iron,* edited by T. A. Wertime and J. D. Muhly. New Haven: Yale University Press, 1980, pp. 267–34.

Tylecote, R. F. *A History of Metallurgy,* 2nd ed. London: Institute of Metals, 1992.

Bronze

Chase, W. T., and T. O. Ziebold. "Ternary Representations of Ancient Chinese Bronze Compositions." In *Archaeological Chemistry II,* edited by G. F. Carter. Washington, D.C.: American Chemical Society, 1978, pp. 292–334.

Chikwendu, V. E., P. T. Craddock, R. M. Farquhar, T. Shaw, and A. C. Umeji. "Nigerian Sources of Copper, Lead and Tin for the Igbo-Ukwu Bronzes." *Archaeometry,* vol. 31 (1989), pp. 27–36.

Craddock, P. T. *Early Metal Mining and Production.* Edinburgh: Edinburgh University Press, 1995.

Hanson, V. F., J. H. Carlson, K. M. Papouchado, and N. A. Nielsen. "The Liberty Bell: Composition of the Famous Failure." *American Scientist,* vol. 14 (1976), pp. 614–19.

Healy, J. F. *Mining and Metallurgy in the Greek and Roman World.* London: Thames and Hudson, 1978.

Tylecote, R. F. *The Prehistory of Metallurgy in the British Isles.* London: The Institute of Metals, 1986.

Tin

Franklin, A. D., J. S. Olin, and T. A. Wertime, eds. *The Search for Ancient Tin.* Washington, D.C.: U.S. Government Printing Office, 1978.

Muhly, J. D. "Tin Trade Routes of the Bronze Age." *American Scientist,* vol. 61 (1973), pp. 404–413.

Penhallurick, R. D. *Tin in Antiquity.* London: Institute of Metals, 1986.

Yener, K. A., M. Goodway, H. Özbal, E. Kaptan, and A. N. Pehlivan. "Kestel: An Early Bronze Age Source of Tin Ore in the Taurus Mountains, Turkey." *Science,* vol. 244 (1989), pp. 200–3.

Lead

Gale N. H., and Z. A. Stos-Gale. "Bronze Age Archaeometallurgy of the Mediterranean: The Impact of Lead Isotope Studies." In *Archaeological Chemistry IV,* edited by R. O. Allen. Washington, D.C.: American Chemical Society, 1989, pp. 159–98.

———. "Lead Isotope Studies in the Aegean." In *New Developments in Archaeological Science,* edited by A. M. Pollard. Oxford: Oxford University Press, 1992, pp. 63–108.

Hong, I. S., J.-P. Candelone, C. C. Patterson, and C. F. Boutron. "Greenland Ice Evidence of Hemispheric Lead Pollution Two Millennia Ago by Greek and Roman Civilizations." *Science,* vol. 265 (1994), pp. 1841–43.

Nriagu, J. O. *Lead and Lead Poisoning in Antiquity.* New York: Wiley-Interscience, 1983.

Waldron, H. A., A. Mackie, and A. Townshend. "The Lead Content of Some Romano-British Bones." *Archaeometry,* vol. 18 (1976), pp. 221–27.

Zinc and Brass

Craddock, P. T. "Medieval Copper Alloy Production and West African Bronze Analyses—Part I." *Archaeometry,* vol. 27 (1985), pp. 17–51.

Hegde, K. T. M., P. T. Craddock, and V. H. Sonavane. "Zinc Distillation in Ancient India." In *Proceedings of the 24th International Archaeometry Symposium,* edited by J. S. Olin and M. J. Blackman. Washington, D.C.: Smithsonian Institution Press, 1986, pp. 249–58.

Michel, H. V., and F. Asaro. "Chemical Study of the Plate of Brass." *Archaeometry,* vol. 21 (1979), pp. 3–19.

Werner, O., and F. Willett. "The Composition of Brasses from Ife and Benin." *Archaeometry,* vol. 17 (1975), pp. 141–56.

Iron and Steel

Schmidt, P. R., and S. T. Childs. "Ancient African Iron Production." *American Scientist,* vol. 83 (1995), pp. 524–33.

Sherby, O. D., and J. Wadsworth. "Damascus Steels." *Scientific American,* vol. 252, no. 2 (1985), pp. 112–20.

Wertime, T. A. *The Coming of the Age of Steel.* Chicago: The University of Chicago Press, 1962.

Wertime, T. A., and J. D. Muhly, eds. *The Coming of the Age of Iron.* New Haven: Yale University Press, 1980.

Gold

Lechtman, H. "Pre-Columbia Surface Metallurgy." *Scientific American,* vol. 250, no. 6 (1984), pp. 56–63.

Oddy, W. A., T. G. Padley, and N. D. Meeks. "Some Unusual Techniques of Gilding in Antiquity." In *Proceedings of the 18th International Symposium on Archaeometry and Archaeological Prospection,* edited by I. Scollar. Cologne: Rheinland-Verlag, 1979, pp. 230–42.

Silver

Boon, G. C. "Counterfeiting in Roman Britain." *Scientific American,* vol. 231, no. 6 (1974), pp. 120–30.

Meyers, P. "Applications of X-Ray Radiography in the Study of Archaeological Objects." In *Archaeological Chemistry II,* edited by G. F. Carter. Washington, D.C.: American Chemical Society. 1978, pp. 79–96.

Meyers, P., L. Van Zelst, and E. V. Sayre. "Major and Trace Elements in Sassanian Silver." In *Archaeological Chemistry,* edited by C. W. Beck. Washington, D.C.: American Chemical Society, 1974, pp. 22–33.

Stós-Fertner, Z., and N. H. Gale. "Chemical and Lead Isotope Analysis of Ancient Egyptian Gold, Silver, and Lead." In *Proceedings of the 18th International Symposium on Archaeometry and Archaeological Prospection,* edited by I. Scollar. Cologne: Rhein-land-Verlag, 1979, pp. 299–314.

Mercury

Goldwater, L. J. *Mercury: A History of Quicksilver.* Baltimore, Md.: York Press, 1972.

Platinum

Kronberg, B. J., L. L. Coatsworth, and M. C. Usselman. "Mass Spectrometry as a Historical Probe: Quantitative Answers to Historical Questions in Metallurgy." In *Archaeological Chemistry III,* edited by J. B. Lambert. Washington, D.C.: American Chemical Society, 1984, pp. 295–310.

Chapter 8: HUMANS

Isotopic Analysis for Diet

Ambrose, S. H. "Isotopic Analysis of Paleodiets: Methodological and Interpretive Considerations." In *Investigations of Ancient Human Tissue,* edited by M. K. Sandford. Langhorne, Pa.: Gordon and Breach, 1993, pp. 59–130.
DeNiro, M. J. "Stable Isotopy and Archaeology." *American Scientist,* vol. 75 (1987), pp. 182–91.
Van der Merwe, N. J. "Carbon Isotopes, Photosynthesis, and Archaeology." *American Scientist,* vol. 70 (1982), pp. 596–606.
———. "Light Stable Isotopes and the Reconstruction of Prehistoric Diets." In *New Developments in Archaeological Science,* edited by A. M. Pollard. Oxford: Oxford University Press, 1992, pp. 247–64.

Elemental Analysis and Diet

Lambert, J. B., S. V. Simpson, C. B. Szpunar, and J. E. Buikstra. "Ancient Human Diet from Inorganic Analysis of Bone." *Accounts of Chemical Research,* vol. 17 (1984), pp. 298–305.
Radosevich, S. C. "The Six Deadly Sins of Trace Element Analysis: A Case of Wishful Thinking in Science." In *Investigations of Ancient Human Tissue,* edited by M. K. Sandford. Langhorne, Pa.: Gordon and Breach, 1993, pp. 269–332.

Toxic Effects of Lead

Aufderheide, A. C., F. D. Neiman, L. E. Wittmers, and G. Rapp. "Lead in Bone II: Skeletal-Lead Content as an Indicator of Lifetime Lead Ingestion and the Social Correlates in an Archaeological Population." *American Journal of Physical Anthropology,* vol. 55 (1981), pp. 285–91.

Corruccini, R. S., A. C. Aufderheide, J. S. Handler, and L. E. Wittmers, Jr. "Patterning of Skeletal Lead Content in Barbados Slaves." *Archaeometry,* vol. 29 (1987), pp. 233–39.

Sandford, M. K. "Understanding the Biogenic-Diagenetic Continuum: Interpreting Elemental Concentrations of Archaeological Bone." In *Investigations of Ancient Human Tissue,* edited by M. K. Sandford. Langhorne, Pa.: Gordon and Breach, 1993, pp. 11–15.

Dating by Elemental Analysis

Haddy, A., and A. Hanson. "Nitrogen and Fluorine Dating of Moundville Skeletal Samples." *Archaeometry,* vol. 24 (1982), pp. 37–44.

Oakley, K. P. "Analytical Methods of Dating Bones." In *Science in Archaeology,* 2nd ed., edited by D. Brothwell and E. Higgs. New York: Praeger, 1970, pp. 35–45.

Thomson, K. S. "Piltdown Man: The Great English Mystery Story." *American Scientist,* vol. 79 (1991), pp. 194–201.

Prehistoric Organic Molecules

Bethell, P. H., R. P. Evershed, and L. J. Goad. "The Investigation of Lipids in Organic Residues by Gas Chromatography/Mass Spectrometry: Applications to Palaeodietary Studies." In *Prehistoric Human Bone: Archaeology at the Molecular Level,* edited by J. B. Lambert and G. Grupe. Berlin: Springer-Verlag, 1993, pp. 227–55.

Callen, E. O. "Diet as Revealed by Coprolites." In *Science in Archaeology,* edited by D. Brothwell and E. Higgs. New York: Praeger, 1970, pp. 235–43.

Evershed, R. P. "Lipids from Samples of Skin from Seven Dutch Bog Bodies: Preliminary Report." *Archaeometry,* vol. 32 (1990), pp. 139–53.

Evershed, R. P., G. Turner-Walker, R. E. M. Hedges, N. Tuross, and A. Leyden. "Preliminary Results for the Analysis of Lipids in Ancient Bone." *Journal of Archaeological Science,* vol. 22 (1995), pp. 277–90.

Loy, T. H. "Prehistoric Organic Residues: Recent Advances in Identification, Dating, and their Antiquity." In *Archaeometry '90,* edited by E. Pernicka and G. A. Wagner. Basel: Birkhäuser Verlag, 1991, pp. 645–656.

Amino Acid Dating

Gillard, R. D., A. M. Pollard, P. A. Sutton, and D. K. Whittaker. "An Improved Method for Age at Death Determination from the Measurement of D-Aspartic Acid in Dental Collagen." *Archaeometry,* vol. 32 (1990), pp. 61–70.

Masters, P. A., and J. L. Bada. "Amino Acid Racemization Dating of Bone and Shell." In *Archaeological Chemistry II,* edited by G. F. Carter. Washington, D.C.: American Chemical Society, 1978, pp. 117–38.

Prior, C. A., P. J. Ennis, E. A. Noltmann, P. E. Hare, and R. E. Taylor. "Variations in D/L Aspartic Acid Ratios in Bones of Similar Age and Temperature History." In *Proceedings of the 24th International Archaeometry Symposium,* edited by J. S. Olin and M. J. Blackman. Washington, D.C.: Smithsonian Institute Press, 1986, pp. 487–98.

Electron Spin Resonance Dating

Grün, R., and C. B. Stringer. "Electron Spin Resonance Dating and the Evolution of Modern Humans." *Archaeometry,* vol. 33 (1991), pp. 153–99.

Sales, K. D., G. V. Robins, and D. Oduwole. "Electron Spin Resonance Study of Bones from the Paleolithic Site of Zhoukoudian, China." In *Archaeological Chemistry IV,* edited by R. O. Allen. Washington, D.C.: American Chemical Society, 1989, pp. 353–68.

Genetic Histories Based on Modern Blood Factors

Ammerman, A. J., and L. L. Cavalli-Sforza. *The Neolithic Transition and the Genetics of Populations in Europe,* Princeton, N.J.: Princeton University Press, 1984.

Cavalli-Sforza, L. L., P. Menozzi, and A. Piazza. *The History and Geography of Human Genes.* Princeton, N.J.: Princeton University Press, 1994.

Menozzi, P., A. Piazza, and L. Cavalli-Sforza. "Synthetic Maps of Human Gene Frequencies in Europeans." *Science,* vol. 201 (1978), pp. 786–792.

Piazza, A. "Blood Genetic Markers: Their Use for Studying Migrations." In *Génétique des Populations Humaines,* edited by E. Ohayon and A. Cambon-Thomasen. Paris: INSERM, 1986, pp. 75–94.

Ancient DNA

Hauswirth, W. W., C. D. Dickel, D. J. Rowold, and M. A. Hauswirth. "Inter- and Intrapopulation Studies of Ancient Humans." *Experientia,* vol. 50 (1994), pp. 585–91.

Herrmann, B., and S. Hummel, eds. *Ancient DNA: Recovery and Analysis of Genetic Material from Paleontological, Archaeological, Museum, Medical, and Forensic Specimens.* Berlin: Springer-Verlag, 1994.

Pääbo, S. "Ancient DNA: Extraction, Characterization, Molecular Cloning, and Enzymatic Amplification." *Proceedings of the National Academy of Sciences of the USA,* vol. 86 (1989), pp. 1939–43.

Powledge, T. M., and M. Rose. "The Great DNA Hunt, Part II: Colonizing the Americas." *Archaeology,* vol. 49, no. 6 (1996), pp. 58–68.

Richards, M. B., B. C. Sykes, and R. E. M. Hedges. "Authenticating DNA Extracted from Ancient Skeletal Remains." *Journal of Archaeological Science,* vol. 22 (1995), pp. 291–99.

Sutton, M. Q., M. Malik, and A. Ogram. "Experiments on the Determination of Gender from Coprolites by DNA Analysis." *Journal of Archaeological Science,* vol. 23 (1996), pp. 263–67.

Modern DNA

Ayala, F. J. "The Myth of Eve: Molecular Biology and Human Origins." *Science,* vol. 270 (1995), pp. 1930–36.

Hammer, M. F. "A Recent Common Ancestry for Human Y Chromosomes." *Nature,* vol. 378 (1995), pp. 376–78.

Horai, S., K. Hayasaka, R. Kondo, K. Tsugane, and N. Takahata. "Recent African Origin of Modern Humans Revealed by Complete Sequence of Hominoid Mitochondrial DNAs." *Proceedings of the National Academy of Sciences of the USA,* vol. 92 (1995), pp. 532–36.

Lewin, R. "The Biochemical Route to Human Origins." *Mosaic,* vol. 22, no. 3 (1991), pp. 46–55.

Wilson, A. C., and R. L. Cann. "The Recent African Genesis of Humans." *Scientific American,* vol. 266 (1992), pp. 68–73.

List of Figure Sources

Figure 1.1 H. Bowman et al., "The Northern Colossus of Memnon: New Slants," *Archaeometry,* vol. 26 (1984), pp. 220.

Figure 1.2 Bowman et al., "Northern Colossus of Memnon," pp. 219.

Figure 1.3 N. Herz and D. B. Wenner, "Tracing the Origins of Marble," *Archaeology,* vol. 34, no. 5 (1981), pp. 16.

Figure 1.4 p. Roos et al., "Chemical, Isotopic and Petrographic Characterization of Ancient White Marble Quarries," in *Proceedings of the 26th International Archaeometry Symposium,* edited by R. M. Farquhar et al. (Toronto: University of Toronto, 1988), p. 221.

Figure 1.5 H. Craig and V. Craig, "Greek Marbles: Determination of Provenance by Isotopic Analysis," *Science,* vol. 176 (1972), p. 401.

Figure 1.6 K. Germann et al., "Determination of Marble Provenance: Limits of Isotopic Analysis," *Archaeometry,* vol. 22 (1980), p. 100.

Figure 1.7 G. de G. Sieveking et al., "Prehistoric Flint Mines and Their Identification as Source of Raw Material," *Archaeometry,* vol. 14 (1972), p. 152.

Figure 1.8 Sieveking et al., "Prehistoric Flint Mines and Their Identification as Source of Raw Material," p. 154.

Figure 1.9 p. F. Healy, H. I. McKillop, and G. B. Walsh, "Analysis of Obsidian from Moho Cay, Belize: New Evidence on Classic Maya Trade Routes," *Science,* vol. 225 (1984), p. 415.

Figure 1.10 N. H. Gale, "Mediterranean Obsidian Source Characterisation by Strontium Isotope Analysis," *Archaeometry,* vol. 23 (1981), p. 48.

Figure 1.11 O. Williams-Thorpe, "Obsidian in the Mediterranean and Near East: A Provenancing Success Story," *Archaeometry,* vol. 37 (1995), p. 227.

Figure 1.12 R. O. Allen and S. E. Pennell, "Rare Earth Element Distribution Patterns for Characterizing Soapstone Artifacts," in *Archaeological Chemistry II,* edited by G. F. Carter (Washington, D.C.: American Chemical Society, 1978), p. 246.

Figure 1.13 O. Williams-Thorpe and R. S. Thorpe, "Millstone Provenancing in Tracing the Route of a Fourth-Century B.C. Greek Merchant Ship," *Archaeometry,* vol. 32 (1990), p. 132.

Figure 1.14 N. Herz et al., "Science versus Art History: The Cleveland Museum Head of Pan and the Miltiades Marathon Victory Monument," *Archaeometry,* vol. 31 (1989), p. 162.

Figure 1.15 M. A. Bello and A. Martín, "Microchemical Characterization of Building Stone from Seville Cathedral, Spain," *Archaeometry,* vol. 34 (1992), p. 22.

Figure 1.16 R. I. Dorn, "Rock Varnish," *American Scientist,* vol. 79 (1991), p. 551.

Figure 1.17 W. E. Kingery, p. B. Vandiver, and M. Prickett, "The Beginnings of Pyro-technology, Part II: Production and Use of Lime and Gypsum Plaster in the Pre-Pottery Neolithic Near East," *Journal of Field Archaeology,* vol. 15 (1988), p. 237.

Figure 2.1 Z. Goffer, in *Archaeological Chemistry* (New York: John Wiley, 1980), p. 334.

Figure 2.2 M. Cowell, *Chemistry in Britain,* (1992), p. 892.

Figure 2.3 A. Sánchez, M. L. Cañabate, and R. Lizcano, "Phosphorus Analysis at Archaeological Sites: An Optimization of the Method and Interpretation of the Results," *Archaeometry,* vol. 38 (1996), p. 160.

Figure 2.4 D. A. Davidson, "Particle Size and Phosphate Analysis--Evidence for the Evolution of a Tell," *Archaeometry,* vol. 25 (1973), p. 145.

Figure 2.5 J. S. Conway, "An Investigation of Soil-Phosphorus Distribution within Occupation Deposits from a Romano-British Hut Group," *Journal of Archaeological Science,* vol. 10, no. 2 (1983), p. 118.

Figure 2.6 Conway, "Investigation of Soil-Phosphorus Distribution," p. 118.

Figure 2.7 L. Biek, "Soil Silhouettes," in *Science in Archaeology,* edited by D. Broth-well and E. Higgs (New York: Praeger, 1970), opposite p. 65.

Figure 3.1 M. F. Kaplan, G. Harbottle, and E. V. Sayre, "Multi-Disciplinary Analysis of Tell el Yahudiyeh Ware," *Archaeometry,* vol. 24 (1982), p. 129.

Figure 3.2 H. W. Catling and A. Millett, "A Study of the Inscribed Stirrup-Jars from Thebes," *Archaeometry,* vol. 8 (1965), p. 6.

Figure 3.3 E. Cooper, *A History of Pottery* (London: Longman Group, 1972), p. 37.

Figure 3.4 Adapted from C. D. McLaughlin, *Development of Photoelectron Spectroscopy as a Method for the Analysis of Archaelogical Artifacts,* Ph.D. diss. (Evanston, Ill.: Northwestern University, 1976).

Figure 3.5 G. C. Nelson, *Ceramics: A Potter's Handbook* (New York: Holt, Rinehart and Winston, 1960), p. 159.

Figure 3.6 Nelson, *Ceramics: A Potter's Handbook,* p. 159.

Figure 3.7 J. S. Olin and M. J. Blackman, "Compositional Classification of Mexican Majolica Ceramics of the Spanish Colonial Period," in *Archaeological Chemistry IV,* edited by R. O. Allen (Washington, D.C.: American Chemical Society, 1989), p. 100.

Figure 3.8 E. C. Joel et al., "Lead Isotope Studies of Spanish-Colonial and Mexican Majolica," in *Proceedings of the 16th International Archaeometry Symposium,* edited by R. M. Farquhar et al. (Toronto: University of Toronto, Archaeometry Labora-tory, 1988), p. 191.

Figure 3.9 R. Reilly, *Wedgwood Jasper* (New York: Thames and Hudson, 1994), plate C57.

Figure 3.10 W. B. R. Neave-Hill, *Chinese Ceramics* (Edinburgh: John Bartholemew, 1975), p. 92.

Figure 4.1 K. Nassau, *The Physics and Chemistry of Color* (New York: John Wiley, 1983), p. 11.

Figure 4.2 W. Noll, R. Holm, and L. Born, "Painting of Ancient Ceramics," *Angewandte Chemie, International Edition (English),* vol. 14 (1975), p. 604.

Figure 4.3 A. A. Hassan and F. A. Hassan, "Source of Galena in Predynastic Egypt at Nagada," *Archaeometry,* vol. 23 (1981), p. 78.

Figure 4.4 Hassan and Hassan, "Source of Galena in Predynastic Egypt at Nagada," p. 80.

Figure 4.5 M. S. Tite, M. Bimson, and M. R. Cowell, "Technological Examination of Egyptian Blue," in *Archaeological Chemistry III,* edited by J. Lambert (Washington, D.C.: American Chemical Society, 1984), p. 217.

Figure 4.6 p. E. McGovern and R. H. Michel, "Royal Purple Dye: The Chemical Reconstruction of the Ancient Mediterranean Industry," *Accounts of Chemical Research,* vol. 23 (1990), p. 153.

Figure 4.7 Provided by author.

Figure 4.8 P. E. McGovern, "Archaeological Chemistry: An Emerging Discipline," *Beckman Center New,* vol. 5, no. 2 (1988), p. 5.

Figure 4.9 G. V. Foster and p. J. Moran, "Plants, Paints, and Pottery: Identification of Madder Pigment on Cypriot Ceramicware," in *Archaeometry,* edited by Y. Maniatis (Amsterdam: Elsevier, 1984), p. 186.

Figure 4.10 K. A. Jakes, J. E. Katon, and p. A. Martoglio, "Identification of Dyes and Characterization of Fibers by Infrared and Visible Microspectroscopy: Application to Paracas Textiles," in *Archaeometry '90,* edited by E. Pernicka and G. A. Wagner (Basel: Birkhäuser Verlag, 1991), p. 314.

Figure 4.11 M. V. Orna et al., "Applications of Infrared Microspectroscopy to Art Historical Questions about Medieval Manuscripts," in *Archaeological Chemistry IV,* edited by R. O. Allen (Washington, D.C.: American Chemical Society, 1989), p. 286.

Figure 4.12 S. Best et al., "A Bible Laid Open," *Chemistry in Britain* (1993), p. 119.

Figure 4.13 M. Járó, E. Gondár, and A. Tóth, "Reconstruction of Gilding Technique Used for Medieval Membrane Threads in Museum Textiles," in *Archaeometry '90,* edited by E. Pernicka and G. A. Wagner (Basel: Birkhäuser Verlag, 1991), p. 318.

Figure 4.14 T. A. Cahill et al., "Gutenberg's Inks and Papers: Non-Distinctive Compositional Analysis by Proton Milliprobe," *Archaeometry,* vol. 26 (1984), p. 12.

Figure 5.1 A. Aspinall et al., "Neutron Activation Analysis of Faience Beads," *Archaeometry,* vol. 14 (1972), plate 1 after p. 30.

Figure 5.2 Aspinall et al., "Neutron Activation Analysis of Faience Beads," p. 36.

Figure 5.3 M. S. Tite, "Characterisation of Early Vitreous Materials," *Archaeometry,* vol. 29 (987), p. 27.

Figure 5.4 F. Neuberg, *Ancient Glass* (London: Barrie & Rockliff, 1962), p. 7.

Figure 5.5 R. W. Smith, "The Analytical Study of Glass in Archaeology," in *Science in Archaeology,* edited by D. Brothwell and E. Higgs (New York: Praeger, 1970), p. 615.

Figure 5.6 Freer Gallery of Art, Smithsonian Institution, Washington, D.C.

Figure 5.7 D. J. Barber and I. C. Freestone, "An Investigation of the Origin of the Colour on the Lycurgus Cup by Analytical Transmission Electron Microscopy," *Archaeometry,* vol. 32 (1990), p. 39.

Figure 5.8 E. V. Sayre, "Summary of the Brookhaven Program of Analysis of Ancient Glass," in *Application of Science in Examination of Works of Art* (Boston: Museum of Fine Arts, 1965), p. 153.

Figure 5.9 J. Henderson, M. Tregear, and N. Wood, "The Technology of Sixteenth- and Seventeenth-Century Chinese *Cloisonné* Enamels," *Archaeometry,* vol. 31 (1980), p. 135.

Figure 5.10 J. Henderson, "The Scientific Analysis of Ancient Glass and Its Archaeological Interpretation," in *Scientific Analysis of Archaeology,* edited by J. Henderson (Oxford: Oxford University Press, 1989), p. 42.

Figure 5.11 R. H. Brill, "Chemical Analyses of Some Glasses from Frattesina," *Journal of Glass Studies,* vol. 34 (1992), p. 16.

Figure 5.12 N. H. Tennent et al., "Major, Minor, and Trace Element Analysis of Medieval Stained Glass by Flame Atomic Absorption Spectroscopy," in *Archaeological Chemistry III,* edited by J. B. Lambert (Washington, D.C.: American Chemical Society, 1984), p. 146.

Figure 6.1 T. Oudemans and S. Boon, "Molecular Archaeology: Analysis of Charred (Food) Remains from Prehistoric Pottery by Pyrolysis-Gas Chromatography/Mass Spectrometry," *Journal of Analytical and Applied Pyrolysis,* vol. 20 (1991), p. 207.

Figure 6.2 C. A. Hastorf and M. J. DeNiro, "Reconstruction of Prehistoric Plant Production and Cooking Practices by a New Isotopic Method," *Nature,* vol. 315 (1985), p. 490.

Figure 6.3 F. S. McLaren, J. Evans, and G. C. Hillman, "Identification of Charred Seeds from Epipodaeolithic Sites of Southwest Asia," in *Archaeometry '90,* edited by E. Pernicka and G. A. Wagner (Basel: Birkhäuser Verlag, 1991), p. 801.

Figure 6.4 G. K. Sams, "Beer in the City of Midas," *Archaeology,* vol. 30 (1977), pp. 108, 112.

Figure 6.5 T. Unwin, *Wine and the Vine* (London: Routledge, 1991), p. 65.

Figure 6.6 M. D. Coe, *Breaking the Maya Code* (New York: Thames and Hudson, 1992), p. 247.

Figure 6.7 E. J. W. Barber, *Prehistoric Textiles* (Princeton, N.J.: Princeton University Press, 1991), p. 40.

Figure 6.8 Barber, *Prehistoric Textiles,* p. 34.

Figure 6.9 Barber, *Prehistoric Textiles,* p. 127.

Figure 6.10 M. Raheel, "History, Identification, and Characterization of Old World Fibers and Dyes," in *Ancient Technologies and Archaeological Materials,* edited by S. U. Wisseman and W. S. Williams (Langhome, Pa.: Gordon and Breach, 1994), pp. 126, 136.

Figure 6.11 A. Betts et al., "Early Cotton in North Africa," *Journal of Archaeological Sciences,* vol. 21, no. 4 (1994), p. 490.

Figure 6.12 Betts et al., "Early Cotton in North Africa," p. 494.

Figure 6.13 R. A. Skelton, T. E. Marston, and C. D. Painter, *The Vinland Map and the Tartar Relation,* 2nd ed. (New Haven: Yale University Press, 1996), frontispiece.

Figure 6.14 N. S. Baer and N. Indictor, "Chemical Investigations of Ancient Near Eastern Archaeological Ivory Artifacts," in *Archaeological Chemistry,* edited by C. W. Beck (Washington, D.C.: American Chemical Society, 1974), p. 241.

Figure 6.15 R. F. Marschner and H. T. Wright, "Asphalts from Middle Eastern Archaeological Sites," in *Archaeological Chemistry II,* edited by G. F. Carter (Washington, D.C.: American Chemical Society, 1978), p. 152.

Figure 6.16 Yorkshire Museum, York, U.K.

Figure 6.17 N. Robinson, et al., "Proof of a Pine Wood Origin for Pitch from Tudor (*Mary Rose*) and Etruscan Shipwrecks, Application of Analytical Organic Chemistry in Archaeology," *Analyst,* vol. 112 (1987), p. 639.

Figure 6.18 K. Evans and C. Heron, "Disinfectant and Chewing Gum: Natural Products Chemistry in Archaeology," *Chemistry & Industry,* (1993), p. 448.

Figure 6.19 C. W. Beck, "Spectroscopic Investigations of Amber," *Applied Spectroscopy Reviews,* vol. 22 (1986), p. 64.

Figure 6.20 J. S. Mills and R. White, *The Organic Chemistry of Museum Objects,* 2nd ed. (Oxford, England: Butterworth-Heinemann, 1994), p. 112.

Figure 6.21 J. Glastrup, "An Easy Identification Method of Waxes and Resins," in *Archaeometry,* edited by Y. Maniatis (Amsterdam: Elsevier, 1984), pp. 247-48.

Figure 7.1 C. S. Smith, "The Interpretation of Microsfunction of Metallic Artifacts," in *Application of Science in Examination of Works of Art,* edited by p. T. Rathbone and W. J. Young (Boston: Museum of Fine Arts, 1967), pp. 29, 30.

Figure 7.2 p. T. Craddock, *Early Metal Mining and Production* (Edinburgh: Edinburgh University Press, 1995), p. 101.

Figure 7.3 G. Rapp, Jr., J. Allert, and E. Henrickson, "Trace Element Discrimination of Discrete Sources of Native Copper," in *Archaeological Chemistry III,* edited by J. Lambert (Washington, D.C.: American Chemical Society, 1984), p. 282.

Figure 7.4 R. G. V. Hancock et al., "Distinguishing European Trade Copper and Northeastern North American Native Copper," *Archaeometry,* vol. 33 (1991), p. 79.

Figure 7.5 p. T. Craddock, *Early Metal Mining and Production* (Edinburgh: Edinburgh University Press, 1995), p. 133.

Figure 7.6 Craddock, *Early Metal Mining and Production,* p. 138.

Figure 7.7 R. F. Tylecote, *The Prehistory of Metallurgy in the British Isles* (London: Institute of Metals, 1986), p. 83.

Figure 7.8 Virginia Museum of Fine Arts, Richmond, Virginia.

Figure 7.9 Museum of Fine Arts, Boston, Massachusetts.

Figure 7.10 K. A. Yener and p. B. Vandiver, "Tin Processing at Göltepe, an Early Bronze Age Site in Anatolia," *American Journal of Archaeology,* vol. 97 (1993), p. 209.

Figure 7.11 S. Hong et al., "Greenland Ice Evidence of Hemispheric Lead Pollution Two Millennia Ago by Greek and Roman Civilizations," *Science,* vol. 265 (1994), p. 1842.

Figure 7.12 N. H. Gale and Z. A. Stos-Gale, "Lead Isotope Studies in the Aegean," in *New Developments in Archaeological Science,* edited by A. M. Pollard (Oxford: Oxford University Press, 1992), p. 74.

Figure 7.13 Gale and Stos-Gale, "Lead Isotope Studies in the Aegean," p. 76.

Figure 7.14 Gale and Stos-Gale, "Lead Isotope Studies in the Aegean," p. 81.

Figure 7.15 C. L. Goucher et al., "Lead Isotope Analyses and Possible Metal Sources for Nigerian 'Bronzes'," in *Archaeological Chemistry II,* edited by G. F. Carter (Washington, D.C.: American Chemical Society, 1978), p. 280.

Figure 7.16 K. T. M. Hegde, p. T. Craddock, and V. H. Sonavane, "Zinc Distillation in Ancient India," in *Proceedings of the 24th International Archaeometry Symposium,* edited by J. S. Olin and M. J. Blackman (Washington, D.C.: Smithsonian Institution Press, 1986), p. 250.

Figure 7.17 Hegde, Craddock, and Sonavane, "Zinc Distillation in Ancient India," p. 254.

Figure 7.18 O. Werner and F. Willett, "The Composition of Brasses from Ife and Benin," *Archaeometry,* vol. 17 (1975), after p. 142.

Figure 7.19 R. F. Tylecote, *A History of Metallurgy,* 2nd ed. (London: Institute of Materials, 1992), p. 53.

Figure 7.20 Craddock, *Early Metal Mining and Production,* p. 272.

Figure 7.21 The Seattle Art Museum, Seattle, Washington.

Figure 7.22 p. Meyers, L. Van Zelst, and E. V. Sayre, "Major Trace Elements in Sassanian Silver," in *Archaeological Chemistry,* edited by C. W. Beck (Washington, D.C.: American Chemical Society, 1974), p. 29.

Figure 7.23 Z. Stós-Fertner and N. H. Gale, "Chemical and Lead Isotope Analysis of Ancient Egyptian Gold, Silver, and Lead," in *Proceedings of the 18th International Symposium on Archaeometry and Archaeological Prospection,* edited by I. Scollar (Cologne: Rheinland Verlag, 1979), p. 306.

Figure 7.24 p. Meyers, "Applications of X-Ray Radiography in the Study of Archaeological Objects," in *Archaeological Chemistry II,* edited by G. F. Carter (Washington, D.C.: American Chemical Society, 1978), p. 86.

Figure 7.25 G. C. Boon, "Counterfeiting in Roman Britain," *Scientific American,* vol. 231, no. 6 (1974), p. 122.

Figure 7.26 A. A. Gordus and J. p. Gordus, "Neutron Activation Analysis of Gold Impurity Leads in Silver Coins and Art Objects," in *Archaeological Chemistry,* edited by C. W. Beck (Washington, D.C.: American Chemical Society, 1974), p. 135.

Figure 8.1 N. J. van der Merwe, in *New Developments in Archaeological Science,* edited by A. M. Pollard (Oxford: Oxford University Press, 1992), p. 255.

Figure 8.2 M. J. DeNiro, "Stable Isotopy and Archaeology," *American Scientist,* vol. 75 (1987), p. 190.

Figure 8.3 A. Haddy and A. Hanson, "Nitrogen and Fluorine Dating of Moundville Skeletal Samples," *Archaeometry,* vol. 234 (1982), p. 40.

Figure 8.4 R. p. Evershed et al., "Preliminary Results for the Analysis of Lipids in Ancient Bone," *Journal of Archaeological Science,* vol. 22 (1995), p. 284.

Figure 8.5 R. p. Evershed, "Lipids from Samples of Skin from Seven Dutch Bog Bodies: Preliminary Report," *Archaeometry,* vol. 32 (1990), p. 146.

Figure 8.6 Evershed, "Lipids from Samples of Skin from Seven Dutch Bog Bodies," p. 151.

Figure 8.7 p. H. Bethell, R. p. Evershed, and L. J. Goad, "The Investigation of Lipids in Organic Residues by Gas Chromatography/Mass Spectrometry: Applications to

Palaeodietary Studies," in *Prehistoric Human Bone: Archaeology at the Molecular Level*, edited by J. B. Lambert and G. Grupe (Berlin: Springer-Verlag, 1993), p. 245.

Figure 8.8 R. D. Gillard et al., "An Improved Method for Age at Death Determination from the Measurement of D-Aspartic Acid in Dental Collagen," *Archaeometry*, vol. 32 (1990), p. 68.

Figure 8.9 K. D. Sales, G. V. Robins, and D. Oduwole, "Electron Spin Resonance Study of Bones from the Paleolithic Site of Zoukoudian, China," in *Archaeological Chemistry IV*, edited by R. O. Allen (Washington, D.C.: American Chemical Society, 1989), p. 361.

Figure 8.10 L. Cavalli-Sforza, p. Menozzi, and A. Piazza, *The History and Geography of Human Genes* (Princeton, N.J.: Princeton University Press, 1994), p. 154.

Figure 8.11 Cavalli-Sforza, Menozzi, and Piazza, *History and Geography of Human Genes*, p. 78.

Figure 8.12 Cavalli-Sforza, Menozzi, and Piazza, *History and Geography of Human Genes*, p. 156.

Figure 8.13 Cavalli-Sforza, Menozzi, and Piazza, *History and Geography of Human Genes*, p. 82.

Figure 8.14 Cavalli-Sforza, Menozzi, and Piazza, *History and Geography of Human Genes*, p. 191.

Figure 8.15 Cavalli-Sforza, Menozzi, and Piazza, *History and Geography of Human Genes*, p. 250.

Figure 8.16 Cavalli-Sforza, Menozzi, and Piazza, *History and Geography of Human Genes*, p. 292.

Figure 8.17 Cavalli-Sforza, Menozzi, and Piazza, *History and Geography of Human Genes*, p. 59.

Figure 8.18 A. Piazza et al., "A Genetic History of Italy," *Annals of Human Genetics*, vol. 52 (1988), pp. 203-13.

Figure 8.19 M. Nei, *Molecular Evolutionary Genetics* (New York: Columbia University Press, 1987), p. 277.

Figure 8.20 R. L. McCann, M. Stoneking, and A. C. Wilson, "Mitochondrial DNA and Human Evolution," *Nature*, vol. 234 (1987), p. 34.

List of Color Insert Sources

Plate 1   G. Harbottle and p. C. Weigard, "Turquoise in Pre-Columbian America," *Scientific American,* vol. 266, no. 2 (1992), p. 84.

Plate 2   R. I. Dorn, "Rock Varnish," *American Scientist,* vol. 79 (1991), p. 544.

Plate 3   U. Hofmann, "The Chemical Basic of Ancient Greek Vase Painting," *Angewandte Chemie, International Edition (English),* vol. 1 (1962), p. 345.

Plate 4   D. Whitehouse, "The Origins of Italian Maiolica," *Archaeology,* vol. 31, no. 2 (1978), p. 48.

Plate 5   W. B. R. Neave-Hill, *Chinese Ceramics* (Edinburgh: John Bartholomew, 1975), p. 107.

Plate 6   M. Saltzman, "Identifying Dyes in Textiles," *American Scientist,* vol. 80 (1992), p. 478.

Plate 7   Mentor-UNESCO, *Egyptian Wall Paintings from Tombs and Temples* (UNESCO, 1962), pl. 28.

Plate 8   p. Rivet, *Maya Cities* (London: Elek Books, 1960), pl. 3.

Plate 9   Los Angeles Museum of Art

Plate 10   J. B. Eklund, "Art Opens Way for Science," *Chemical and Engineering News* (June 5, 1978), p. 26. Photograph by J. A. Schaffer.

Plate 11   C. Desroches-Noblecourt, *Tutankhamen* (New York: New York Graphic Society, 1963), p. 114.

Plate 12   D. Grose, "The Formation of the Roman Glass Industry," *Archaeology,* vol. 36, no. 4 (1983), p. 40.

Plate 13   D. B. Harden et al., *Glass of the Caesars* (Milan: Olivetti, 1987), p. 61.

Plate 14   Harden et al., *Glass of the Caesars,* p. 246.

Plate 15   R. D. Penhallurick, *Tin in Antiquity* (London: Institute of Metals, 1986), frontispiece and p. 67.

Plate 16   Museo Oro del Peru.

INDEX